Waterflooding: Surveillance and Remediation

Society of Petroleum Engineers

Richardson, Texas, USA

Disclaimer

This book was prepared by members of the Society of Petroleum Engineers and their well-qualified colleagues from material published in the recognized technical literature and from their own individual experience and expertise. While the material presented is believed to be based on sound technical knowledge, neither the Society of Petroleum Engineers nor any of the authors or editors herein provide a warranty either expressed or implied in its application. Correspondingly, the discussion of materials, methods, or techniques that may be covered by letters patents implies no freedom to use such materials, methods, or techniques without permission through appropriate licensing. Nothing described within this book should be construed to lessen the need to apply sound engineering judgment nor to carefully apply accepted engineering practices in the design, implementation, or application of the techniques described herein.

ISBN: 978-1-61399-806-9 [Print]
ISBN: 978-1-61399-807-6 [Mobi (Amazon)]
ISBN: 978-1-61399-808-3 [Epub (iTunes)]
ISBN: 978-1-61399-809-0 [WebPDF (ADE)]

10 9 8 7 6 5 4 3 2 1

Society of Petroleum Engineers
222 Palisades Creek Drive
Richardson, TX 75080-2040 USA

http://store.spe.org
service@spe.org
1.972.952.9393

I would like to thank Royal Dutch Shell for the opportunity to perform such a fascinating role, which enabled me to build a wide skill set in waterflooding and to address waterflood issues in fields all over the world. My thanks also go to the many colleagues and associates who, over the years, have supported, steered, and guided me. Without them, I would know nothing.
—Dave Chappell

Table of Contents

WATERFLOODING: SURVEILLANCE AND REMEDIATION

Dave Chappell

Dave Chappell has spent his career working on waterflood developments and operations in Brunei, Oman, Thailand, and Australia. In 2003, he became one of the founding members of Shell's central waterflood team tasked with improving waterflood performance across the entire Shell waterflood portfolio, based in The Hague, The Netherlands. He went on to manage that group from 2008 until his retirement in 2018. Since then, he has worked as an independent consultant in the waterflood arena.

1. Background

Historically, waterflood has been viewed as largely the responsibility of the reservoir engineer. That was primarily because waterflood considerations tended to be heavily focused on the displacement process and the associated recovery impacts. While those are critically important to waterflood success, it is now much more widely recognized that waterflood is a process with many different moving parts, and it therefore requires the input of a wide range of disciplines, each of which needs to interface effectively with the others, to deliver a fully optimized project.

Waterflood remains by far the most widely used process that uses an external energy source to improve recovery. Furthermore, it has been successfully used for more than a century. The theoretical basis for waterflood displacement has been understood for quite some time and is well-covered in the available literature (Willhite 1986; Warner 2015). Despite the focus on this aspect of waterflood, it is only in recent years that there has been any detailed inspection of the critical success factors across the full range of disciplines involved. Furthermore, not all the factors have been particularly well-documented. This book is one in a series of publications that aims to redress some of those shortcomings by looking at a range of factors influencing success in waterflood design and operation.

2. Introduction

The primary function of well and reservoir surveillance is to ensure the asset is operated in an effective manner to optimize recovery and/or the asset's value. Waterfloods rarely, if ever, proceed in exactly the manner predicted at the time of project sanctioning. This usually occurs because the understanding of the reservoir geology is incomplete. Furthermore, the geological complexity is often underestimated, and this tends to result in a field performance that is poorer than expected.

It is fortunate that there are usually opportunities to modify the flood, allowing any shortcomings to be addressed. Sometimes, the improvements result in a field performance that is even better than originally expected. However, to remediate and improve poor flood performance, it is vital to understand exactly how and where it has gone wrong. Only then can an optimized remediation program be implemented.

This is why facilities, well, and reservoir management should be considered a fundamental part of waterflood management. However, while all projects are likely to benefit from surveillance to identify areas for improvement, some assets are more likely to benefit from this approach than others. In very benign geological settings, it is likely that a reasonably efficient flood will be achieved without any significant intervention, leaving relatively modest volumes as a target for remediation activities. By contrast, the recovery achieved without intervention in complex geological environments is likely to be much more modest, leaving a much larger target for remediation activities. This suggests that the reward available for surveillance and remediation activities is likely to become larger as the complexity of the reservoir increases (**Fig. 1**).

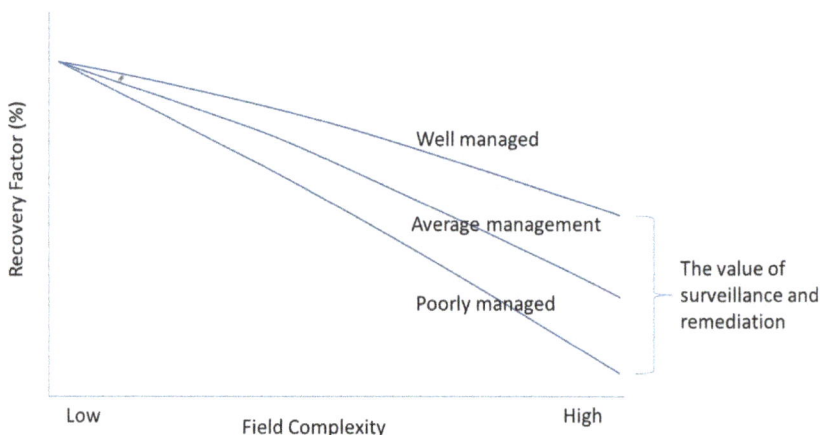

Fig. 1—Value of surveillance and remediation in waterfloods.

One of the primary functions of the surveillance plan is to assess whether the waterflood is sweeping oil in the manner intended at the time of the initial field development plan. A suite of supplementary surveillance elements will also arise from the injection of water, of which some could have a direct impact on the flood itself. For example, when scaling risks arise from the injection of water, problems induced at the producers could materially reduce productivity and there could therefore be associated recovery impacts.

This book describes the various elements that could be included as components in a surveillance plan related to waterflood management. It subsequently looks at the options available to improve flood performance. There will inevitably be surveillance elements required that are not related to the waterflood. Those elements will not be discussed here, and the focus will be entirely on waterflood-related surveillance.

3. Surveillance

3.1 Annual Surveillance Plan. The status of the waterflood will change as the field matures. This means that surveillance requirements could also change over time. It might be found that some elements of the plan can be dropped after some time, either because an element of uncertainty has been removed by the surveillance or because a specific area of concern is no longer an issue (e.g., surveillance for scale squeeze retreatment timing). Because the flood will be expected to change in the medium term, an annual surveillance plan should be compiled.

The program should specify the following data:

- The parameters to be assessed.
- The locations at which those parameters will be measured. This might sometimes require compromises to be made. For example, injection-water quality would ideally be monitored at the bottom of the injection wells. However, this is impractical for ongoing measurement, so samples for this purpose would normally be taken as far downstream as possible. That location will be dependent on the system. In some cases, measurement could be possible at injector wellheads, but in subsea systems a topside measurement must be accepted.
- The frequency with which those parameters will be measured.
- The techniques by which the parameters will be measured. Online continuous metering might enable reductions in manpower requirements, but care is needed to ensure such online analysis has the required accuracy, and attention should always be paid to maintenance and calibration requirements. Manual analyses should always be available as a backup where online analysis is included in the program.
- The resources responsible for sampling and analysis.
- The data storage requirements.
- The resources responsible for data analysis and interpretation. (There is no point spending money on data collection if those data are not used in some way.)

3.2 Surface Facilities Monitoring. Water quality monitoring is required to

- Ensure that chemical injection programs are not only effective but are also fully optimized
- Facilitate the evaluation of the performance of topside equipment
- Ensure that injection-water quality meets the required specifications

3.2.1 Water Sampling. The sampling techniques used greatly influence the validity of the water quality data, particularly

- Sample-point location and assemblies
- Sample taking
- Sample containers
- Sample preservation

Sample quills that penetrate the pipe wall are preferred, enabling samples to be drawn from the center of the pipe. Ideally, the quill should be positioned to face upstream. Many systems, however, only have sidewall sample connections available, with no quill assembly, so samples are taken from the laminar flow layer at the pipe wall. Such sampling points could potentially give erroneously high indications of bacteria and corrosion products. For sidewall sample points, those at the six o'clock position could give results that are artificially high with respect to suspended solids. Therefore, three o'clock positions are preferred.

When samples are taken, the sample points must be adequately flushed first. After flushing, the sample valve should not be adjusted, if possible, because additional loose deposits could break away and contaminate the sample. This is especially important for particle counts and total suspended solids analysis.

Sample bottles must be clean and flushed with sample water before final filling and sealing, except where the bottle contains a preservative (e.g., acid) or when sampling for oil-in-water or bacterial content. Sample bottles should be identified with a number, date, time, and sample point.

When analyzing for pH and dissolved gases, these parameters must be determined in the field because equilibration processes with the atmosphere will otherwise occur, resulting in inaccurate measurements. Changes in these constituents will also subsequently affect determinations for calcium, total hardness, and alkalinity.

Where samples are to be analyzed later, it is likely to be necessary to preserve the samples. The preservation technique will be linked to the analysis that will be performed. For most analyses, the sample can be preserved by adding hydrochloric acid to a pH of 2. When the sample is to be analyzed for ammonia or nitrate, sulfuric acid should be used to acidify it to pH 1.5. When analyzing for volatile fatty acids, sodium hydroxide should be used to increase the pH to greater than 8.5.

In some cases, classical analytical methodologies are used for the analysis of components in injection water (*API RP 45*). However, in many cases, these are being increasingly replaced by more automated technologies such as inductively coupled plasma spectroscopy.

3.2.2 Bacterial Surveillance. Bacterial monitoring is a vital requirement to facilitate the optimization of bacterial control programs. Most routine surveillance programs entail planktonic measurements (enumeration of free-floating bacterial numbers), sessile measurements (bacterial populations attached to surfaces), or a combination of the two. As a result of bacterial growth and activity, the bacteria tend to accumulate in a biofilm constituted of bacterial cells bound together by extracellular bacterial secretions. The organic nature of the biofilm tends to protect the bacteria embedded within it from the biocides that are introduced into the process with the intention of controlling their activity. This means that much higher concentrations of biocide will likely be required to control sessile bacterial populations than are needed to control free-floating planktonic bacteria. It is possible, therefore, that analysis of only planktonic bacterial populations could give misleading information.

It is generally the sessile bacterial populations that cause problems in water-injection systems. However, because planktonic measurements are the easiest to perform, these are often used to infer the extent of sessile populations. It is important to recognize that planktonic counts can confirm only a dirty system, not a clean one. Sustained low numbers in regular planktonic measurements can be used to reduce the frequency of sessile measurements.

Bacterial surveillance is primarily focused on the analysis of sulfate-reducing bacteria (SRB), but general aerobic bacteria (GAB) and general anaerobic bacteria (GAnB) are also important. Unfortunately, SRB have a number of characteristics that make the accurate determination of bacterial numbers difficult:

- They are strictly anaerobic and are inactivated by oxygen.
- They require specialized media, and not all SRB will grow on standard media.
- They can grow slowly and are typically cultured for 30 days, making real-time decisions difficult.

In sampling for bacterial analysis, sample lines should be as short as possible and continuously flowing, or flowed for an adequate time to flush the contents completely. Air must be excluded from the sample for SRB and GAnB measurements, but this is not required for GAB measurement. Sterile sample bottles should be used.

When taking samples for microbiological analysis, it is important that all sampling equipment (such as sample bottles, filters, pipettes) be sterilized to avoid contamination. It is preferable that microbiological testing be performed as soon as possible after sampling to avoid changes to bacterial numbers over time that could give erroneous results. When sampling for SRB, the bottle should be completely filled and closed with a metal screw cap that has a rubber liner.

When samples are taken for the measurement of planktonic bacteria, the water is inoculated into a growth medium. The choice of growth medium is critical to achieving accurate bacterial counts because some bacterial species might grow in one growth medium but not in another. This is a problem for classical bacterial surveillance analyses because it is possible, or even likely, that less than 10% of the bacteria living in a given sample will grow effectively in the given culture medium. Because some media will be more effective than others, this should be assessed for each system.

The most commonly used SRB culture medium is the modified Postgate's B medium. The modification is that the salinity is adjusted to reflect the salinity in the environment being assessed for SRB activity. Most commercially available media use lactate as a carbon source. In oilfield operations, volatile fatty acids are more likely to be used as a carbon source, and media have been developed to take this into account. Typically, the media are liquid, although solid media (agar) are sometimes also used. Agar media are impractical for measuring SRB and GAnB on-site because oxygen must be excluded during growth.

The most commonly used methodology for the measurement of planktonic bacteria is the serial dilution technique (Whitesell 1961). In this method, a sample of water is taken and injected into a bacterial culture broth. After mixing, 1 mL of this sample is injected into another vial containing 9 mL of broth. The process is then repeated in a serial dilution (**Fig. 2**). The broths are then incubated. At the end of

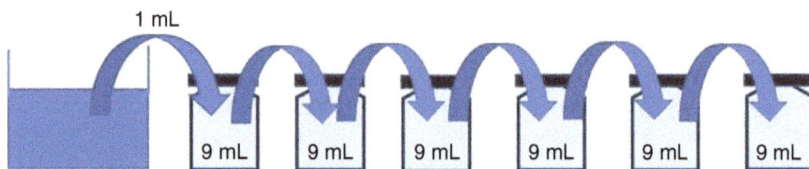

Fig. 2—Serial dilution.

the incubation period, the vials are inspected for the presence of bacteria. In the case of SRB, the presence of these bacteria is indicated by a black coloration in the broth. The serial dilution enables the quantification of the number of bacteria present within the original sample, to within one order of magnitude. In general, counts of 10^2–10^3 per mL are indicative of a system with good bacterial control.

Because this technique only quantifies bacterial populations to within one order of magnitude, it is often augmented by the most probable number (MPN) technique. This technique inoculates 3, 5, or 10 tubes at each serial dilution, and the number of positive responses is noted. The accuracy of the outcome is improved when more tubes are inoculated at each dilution. Based on the number of positive outcomes at each dilution, the most probable number of bacteria in the original sample is inferred from MPN tables.

Because planktonic measurements represent a snapshot of the bacteria present at any given time, they should not be used in isolation to infer the condition of the injection system. When using these data, it is evident that trends in populations are far more important.

A number of possibilities are available for the quantification of sessile bacterial populations:

- Use of scrapings taken from standard flush-mounted corrosion coupons or pipe walls. This can give an indication of SRB numbers if the sample is taken from a known surface area.
- Use of specially designed in-line flush-mounted microbial probes inserted by means of a standard high-pressure access fitting. These probes are usually a modification of commercially available corrosion coupons. They allow for the examination of metal studs, the surfaces of which have been exposed as near as possible to the conditions found on the internal pipe walls. The design of the probe is such that the studs fit flush with the inner wall of the pipeline. The probe can be pulled at any time from the live pipeline, although manipulation requires expertise and special tools. When removed, the studs can be extracted and any bacterial growth studied.
- Use of a sidestream device attached to a low- or high-pressure line from the system. A number of sidestream devices have been designed for use in oilfield systems. One of the most frequently used is a Robbins device. This consists of acrylic plastic tubing for low pressure or a pressure-rated stainless-steel tube for higher pressures. In either case, the tube is fitted with biostuds through the wall. These studs, which are flush mounted, are exposed to the interior of the tubing. They are made of the same material as the pipeline in the system being evaluated (usually carbon steel). Connecting the side stream to a suitable

sampling point and adjusting the water flow to equal the actual flow rate through the pipeline will, in principle, enable the creation of a biofilm on the studs that will be representative of the bacterial layer existing on the pipe wall in the system.

Probes and sidestream devices have the advantage that entire steel biostuds can be assessed for SRB numbers or activity on a routine basis. This provides trends similar to planktonic counts. In addition, the entire biofilm is collected relatively undisturbed.

The procedure for the removal of studs for both a sidestream device and a biological probe is the same. The studs are never touched by hand. They should be transferred to a tube containing growth medium as quickly as possible. The tube is immersed in an ultrasonic bath for sonication to remove the sessile bacteria. The sonication time should be chosen carefully to achieve an effective dispersion of the biofilm while avoiding inactivation of the bacteria. After sonication, 1 mL is withdrawn from the tube using a sterile disposable syringe plunged through the septum. The serial dilution method is then used for bacterial enumeration.

Many assets rely entirely on planktonic bacterial measurements, but there are a number of limitations associated with such a reliance (Maxwell et al. 2002). A primary limitation is that these free-flowing individual bacteria are not the ones causing the real problems in injection systems. Maxwell et al. (2002) conducted routine planktonic measurements downstream of the de-aerators in a system where low numbers of planktonic SRB ($0.9/cm^3$) were detected. This was initially inferred to indicate acceptable process control. However, a corrosion coupon retrieved from the low-pressure header indicated a population of 1.3×10^7 SRB per mL, indicating a high level of biofouling. Subsequent planktonic measurements upstream of the de-aeration process showed that the seawater entering the de-aerators contained less than 1 SRB cell per 1000 mL. This information linked to the downstream measurements showed that significant SRB growth was occurring within the dehydration vessels. The results highlight the value of sessile measurements but also show that planktonic measurements should be conducted throughout the process and that trends in data are important for the interpretation of these data.

Despite advances in growth media and cultivation techniques, it is still acknowledged that only between 1 and 10% of the viable bacteria will be cultured by classical microbiology methods (Maxwell et al. 2004). New molecular methods have emerged based on the measurement of genetic material (DNA and RNA) from bacteria (Skovhus et al. 2007). These techniques, which could require the assistance of specialized laboratories, are now well-developed. They could be useful for specialized problems or periodic surveys of the system. These techniques include

- Direct bacterial counts: Bacterial cells in a dilute suspension are collected on a filter stained with a fluorescent dye, 4-6-diamidino-2-phenylindole. After washing to remove excess dye, the bacterial cells are counted directly with an epifluorescence microscope. This enables the quantification of all microorganisms (living, inactive, and dead) in a sample.
- Fluorescent in-situ hybridization (FISH): The DNA and RNA inside bacteria are unique for each species. Using the sequence of base pairs in the 16S subunit

of ribosomal RNA (rRNA) in bacteria, gene probes can be designed to target different groups of bacteria or individual bacterial species. FISH is a microscopic method in which only the living and active cells are stained with a fluorescent dye that is then visible using epifluorescence microscopy.

- Polymerase chain reaction (PCR): Bacterial DNA is extracted from a sample and then the PCR enables exponential multiplication of DNA, increasing its concentration to measurable amounts (> 10^9 gene copies). With a few adjustments, PCR can be made quantitative, and this is termed real-time quantitative PCR (qPCR). The qPCR assay has a high sensitivity, with a detection limit as low as just a few cells in the initial sample. This technique again captures both living and dead micro-organisms but can be used to target specific species. The amplification techniques now available facilitate very detailed analysis of bacterial populations. Pyrosequencing of 16S rRNA genes was used after a two-step PCR amplification to characterize samples from pigging operations on North Sea platforms to assess impacts related to microbially related corrosion (Mand et al. 2014). Significant SRB counts were recorded. The samples had high linear polarization resistance (LPR) but low weight loss, with corrosion rates suggesting that these methods were not good indicators of corrosion risk. Analysis of the microbial community composition indicated that SRB of the genus *Desulfovibrio* were a significant component, as were sulfide-oxidizing bacteria. This indicated sulfur was a significant intermediate, which would indicate an elevated corrosion risk. Such methods are becoming ever more diagnostic. The increased sophistication of the analysis is resulting in very detailed information about the species present in a given environment, which could be very useful in a detailed quantification of the types of problems that could be present in a system.
- Denaturant gradient gel electrophoresis (DGGE): A sample containing a number of different bacterial strains gives a corresponding number of bands on a DGGE gel. The bands can be cut out of the gel and the organisms identified by means of DNA sequencing. This rapid methodology is often used before other methods (like PCR), when an overview of the bacterial populations is required. DGGE is essentially a bacterial fingerprint that enables profiling of the different bacterial species present in a sample.
- Microautoradiography (MAR): The use of radioactive labeled substrates in combination with MAR allows for the analysis of the metabolic activity of bacteria by the direct visualization of micro-organisms with active substrate uptake. It can be used with microelectrodes or simple staining techniques for the characterization of microbial communities.

Larsen et al. (2005), in examining pigging samples from the Halfdan Field, suggested that molecular surveillance techniques measured at least 100 times greater counts than those obtained using classical techniques. In addition, the new cultivation-independent techniques showed that SRB constituted up to 10% of all bacteria present, despite these being undetectable by classical methods. Furthermore, the new techniques could be applied much more quickly (within a few hours to a few days) compared to traditional techniques (30-day incubation time), enabling a faster response time to losses of process control.

The value of molecular methods has been assessed against traditional surveillance methods in evaluating the performance of different biocides. Bennet and Hoffmann (2018) demonstrated that the molecular methods provided far greater insights into how different bacteria reacted to biocide treatments compared to traditional MPN techniques. The increased granularity of data available from molecular methods highlights the possibility that a reliance on traditional techniques could result in the selection of a suboptimal biocide and treatment regime based on an erroneous belief that the selected control philosophy is effective.

Other chemical analyses that could be relevant to biological activity within a system include

- Total organic carbon: This is a standard test of water quality used to measure the degree of contamination with hydrocarbons. There are a number of standard analytical methods depending on the range, accuracy, and precision required, but in general, the carbon in the water is oxidized to carbon dioxide (CO_2), which is then measured quantitatively.
- Chemical oxygen demand (COD): The basis for this test is that nearly all organic compounds can be fully oxidized to CO_2 with a strong reducing agent under acidic conditions. Therefore, it is effectively a guide to the amount of organic material present in the sample. However, oxygen can also be consumed through the oxidation of inorganic compounds such as ammonia and nitrite. COD is measured as a standardized laboratory assay in which a water sample is incubated with a strong chemical oxidant such as potassium dichromate ($K_2Cr_2O_7$), which is used in combination with boiling sulfuric acid. Because this chemical oxidant is not specific to organic or inorganic oxygen-consuming chemicals, both sources of oxygen demand are measured in a COD assay.
- Biological oxygen demand (BOD): This is the amount of dissolved oxygen needed by aerobic organisms to break down organic material present in a given water sample at a certain temperature over a specific time period. The BOD value is most commonly expressed in milligrams of oxygen consumed per liter of sample during 5 days of incubation at 20°C. Both BOD and COD measure the amount of organic compounds in water, but COD analysis is less specific because it measures everything that can be chemically oxidized rather than just the levels of biologically oxidized organic matter.

It is important to understand that COD and BOD do not necessarily measure the same types of oxygen consumption. For example, COD does not measure the oxygen-consuming potential associated with certain dissolved organic compounds, such as acetate. However, acetate can be metabolized by micro-organisms and would therefore be detected in a BOD assay. Conversely, the oxygen-consuming potential of cellulose is not measured during a short-term BOD assay, but it is measured during a COD test.

3.2.3 Water Quality Monitoring. The most important parameters for water quality and its impact on injection wells are

- Suspended solids loading
- Suspended oil loading (for produced-water-injection systems)

- Bacterial loading (which was covered in Section 3.2.2)
- Oxygen concentration

These govern the amount of corrosion problems that might occur.

Suspended solids are usually measured manually, although online meters are available. If manual sampling is undertaken by operations staff, they could be busy with other, more urgent jobs during process upsets. They might not have the time available to take a routine sample until after the upset has passed, and as a direct consequence, fluctuations in water quality could be missed. Ideally, therefore, sampling and analysis should be performed by a dedicated field chemist.

Total suspended solids measurement is the most commonly used surveillance method. It involves filtering a known volume of water through a filter (0.45 μm is the standard) and weighing the volume of water collecting on the filter. Such analysis is, at best, usually only collected on a daily, or twice-daily, basis. Therefore, online devices allow a better operational reaction to upsets.

Ideally, water quality would be monitored at the bottom of the injection wells. However, this is impractical for regular surveillance, so the wellhead is often the last point at which routine monitoring can occur. The location and number of wells could render this impractical, so routine water quality monitoring could be restricted to key points within the process plant, with occasional measurements taken at wellheads. Some projects use turbidity measurement (sample cloudiness) to infer the suspended solids content.

The sampling point must be selected carefully. Sampling at the bottom of a line would potentially result in elevated solids loading being reported, for example. Ideally, the sample point should be located at the three o'clock position in a vertical section of pipe.

Oil-in-water is also usually measured manually, although, again, online meters are available. There is no standard methodology available for the evaluation of this parameter, and a number of options are available. These include gravimetric analysis, gas chromatography, ultraviolet absorption, fluorescence spectroscopy, and infrared (IR) absorption.

The IR absorption method is the one that is the most commonly applied. The sample is initially acidified if it is not to be analyzed immediately. This is followed by an extraction of hydrocarbons from the water with a solvent and filtration to remove solids and water droplets. The sample is analyzed for IR absorbency at appropriate wavelengths to capture the maximum absorbances of the aliphatic and aromatic carbon-hydrogen bond stretching frequencies. The three-wavelength method records absorption at 3030 cm^{-1} (C-H stretch aromatic vibration), 2960 cm^{-1} (CH-H stretch methylene vibration), and 2930 cm^{-1} (CH$_2$-H stretch methyl vibration). Absorption in samples is then referenced to absorption in a reference oil. Thus, it can be seen that this measurement is not an absolute quantification of the oil-in-water concentration. Furthermore, it is likely that this methodology will extract not only suspended oil but also some oil that is dissolved, rather than dispersed, in the water.

The online oil-in-water meters are beginning to be used more widely. Again, a number of different measurement principles are used by the different commercial units available. Reliability has been a significant problem with such equipment in the past, but this is starting to improve. The use of online cleaning has been critical in improving equipment reliability because the sensors tend to be easily blocked. This is usually achieved through ultrasonic cleaning. Without this, experience suggests that reliability could be lost within hours.

To control corrosion in water-injection systems, most projects specify that oxygen levels be maintained at a level of less than 10 parts per billion (ppb). However, some companies routinely operate at higher oxygen levels (up to 30 ppb) in the belief that the creation of a small corrosion layer will act to passivate additional corrosion at these levels. Maintaining specification is important because when it is lost there will be increased corrosion of the equipment and the corrosion products will increase the solids loading of the injection water.

A number of different methodologies are available for measurement, but each has associated limitations. Ideally, a combination of techniques would be used to identify when specifications are not being met. Simple tube test kits are commonly applied for manual testing, and these are available in different ranges based on the oxygen content of the sample. It is important that tubes of an appropriate range be used so that the oxygen concentration can be easily detected. A number of different colorimetric reagents are available for such measurements, but they all use oxidation-reduction indicators that are converted from a colorless, reduced form to an oxidized, colored form, with the depth of coloration being proportional to the concentration of oxygen. These methods are able to generate results very quickly but, because they are based on oxidation-reduction reactions, the presence of other redox species in the water can influence the accuracy of the results. For example, the oxygen scavenger used in the process is known to interfere with the dye in the tubes. In a well-operated system, the oxygen scavenger should be optimized to minimize the excess oxygen scavenger used. However, it is important to recognize the potential impact this can have on the colorimetric analysis.

Because tube measurements will usually only be taken once or twice a day, it is important that they be supplemented with online metering. This could entail the use of both online oxygen probes and online corrosion probes. The reliability of oxygen probes can be a problem, requiring a regular probe cleaning program.

A number of corrosion probe types can be used, including electrical resistance (ER) probes and LPR probes. LPR probes are often used in conjunction with galvanic oxygen probes, which incorporate two elements, one brass and the other steel, which are exposed to the injected water. A galvanic current is measured between the elements. It will be proportional to the electropotential difference and conductivity of the water. The probe can be supplied with flush or protruding electrode elements. The flush element is usually selected for high flow rates and long life, although a protruding electrode probe provides an improved response.

Where standard 2-in. access fittings are used, it is possible to change between ER and LPR probes on the basis of operational experience. Probes should be hard wired into the distributed control system and continually monitored.

Another important element of surveillance related to oxygen, in cases where a chemical oxygen scavenger is used, is to monitor the amount of residual sulfite. This allows a robust optimization of the chemical-oxygen-scavenger injection. A slight excess of this chemical can be injected, enabling efficient oxygen removal while avoiding any excess injection, which becomes a waste of expenditure. This can have further importance because it has been claimed that excess oxygen scavenger has the potential to induce additional corrosion problems. However, it is likely that any increases in the corrosion rate will be modest. Fig. 3 illustrates a series of such measurements. In this case, they clearly indicate inadequate levels of control of the oxygen-scavenger injection rate.

Residual Sulfite Measurements

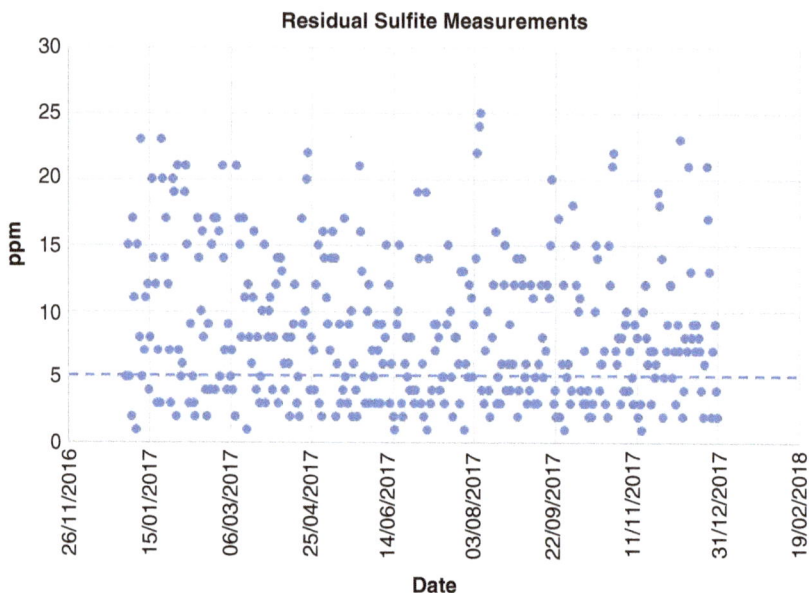

Fig. 3—Residual sulfite measurements. ppm = parts per million.

3.2.4 Compositional Analysis. Compositional data on produced water can provide vitally important information for flood management. One of the most common uses of these data is to manage scaling issues in producing wells arising from the injection of water sources containing sulfate. Seawater, the most commonly used water source in offshore waterflood operations, contains appreciable concentrations of sulfate ions, and sulfate scaling can cause significant problems in such settings (see *Waterflooding: Chemistry*, another book in this series). Collecting compositional data on a regular basis can facilitate an improved understanding of the flow pathways through the reservoir. It can also help to ascertain where scale deposition is occurring and define when scale squeeze treatments are required.

The scale control implemented at the Alba Field (Paulo et al. 2001) provides a good example. **Fig. 4** shows the barium concentrations in the produced water for one of the wells in this field as a function of the seawater fraction in the produced water (as could be quantified from the compositional data). In this field, there was a compositional variation between the oil leg, where barium concentrations were 50 parts per million (ppm), and the aquifer water, where barium concentrations were 80 ppm. The analysis results suggest that before seawater breakthrough, an increasing proportion of the produced water was coming from the aquifer. This was evidenced by the measured barium levels, which increased from 50 to 60 ppm to the aquifer-water concentration of 80 ppm. This implies that, as the waterflood proceeded, an increasing proportion of the produced water was aquifer brine and suggests that the injected water was sweeping through the aquifer, presumably leaving parts of the oil-bearing reservoir unswept.

It was also observed that barium levels dropped approximately 6 weeks before seawater breakthrough. This could potentially have resulted from barium sulfate scaling deposition away from the producing wellbore. In this case, it allowed the

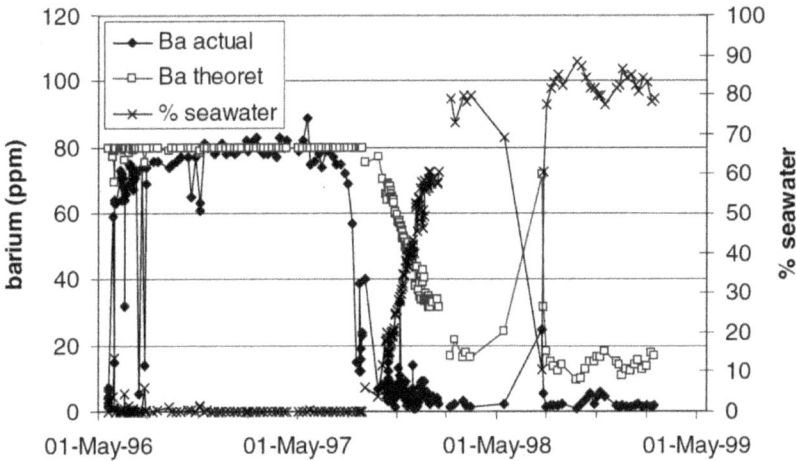

Fig. 4—Comparison of barium (Ba) concentrations and seawater fraction (Paulo et al. 2001).

measurement of barium concentrations to be used as a predictor of the imminent onset of seawater breakthrough.

In another part of the field, a different pattern emerged, with very low barium levels being observed for prolonged periods. This suggests that much more extensive mixing of injected seawater and formation water was occurring within the reservoir. The compositional data, which would have been collected to understand the scaling-deposition risks in any case, were thus also useful in elucidating flow pathways. These appeared to travel primarily through the oil leg in some parts, but in other locations, they revealed flow through the aquifer, which presumably occurred because of gravity slumping through the high-permeability reservoir.

The interpretation of the ion profiles of a number of produced-water samples in terms of the number of contributing water sources and their compositions can be greatly improved by using statistical methods such as principal component analysis (PCA) (Webb and Kuhn 2004). PCA is able to represent the variability in a data set containing many variables with a few pseudovariables that are a combination of the original variables. A single pseudovariable based on all the major ions typically found in produced water can provide a simple, linear measure of the amount of injection-water breakthrough.

With PCA, the onset of injection-water breakthrough could potentially be established with greater sensitivity. **Fig. 5** shows two principal components for four producing wells in the Bittern Field with the respective formation-water/injection-water mixing lines. One of the components represents a linear measure of the extent of water breakthrough, and the other represents the variation of produced-water composition not attributable to water breakthrough (e.g., variation resulting from reservoir heterogeneity). This showed that injection-water breakthrough had not occurred in any wells. As a result, a decision was made to postpone planned scale squeeze treatments.

In another example, seawater breakthrough needed to be understood because of a significant barium sulfate and strontium sulfate scaling risk. One well showed contradictory trends in ion compositional data and where barium and sulfate ion data suggested seawater breakthrough had occurred (**Fig. 6**) (Scheck and Ross 2008).

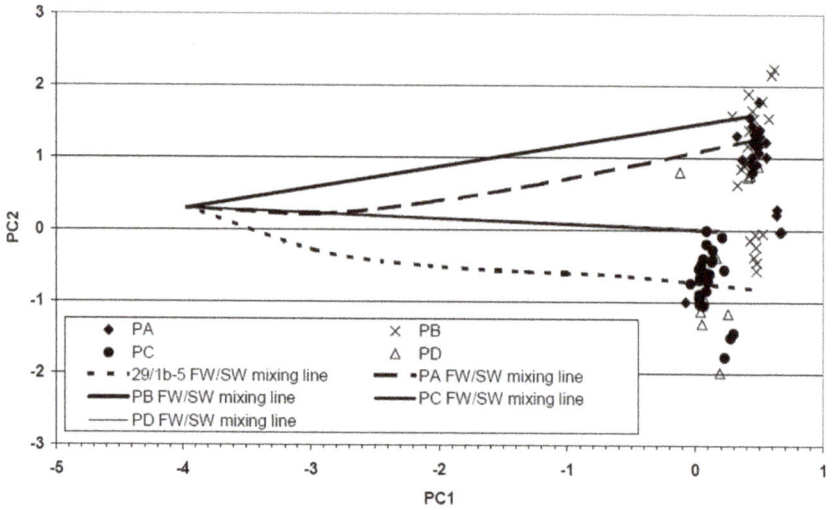

Fig. 5—Principal component analysis on Bittern wells (Webb and Kuhn 2004).

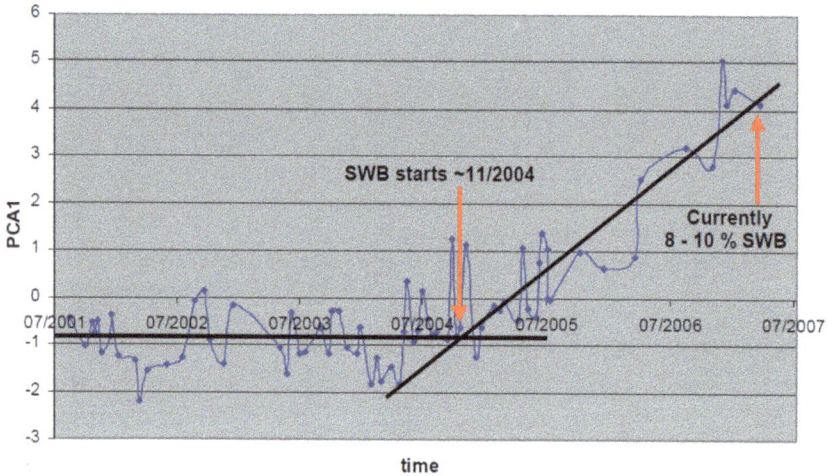

Fig. 6—PCA1 vs. time, showing seawater breakthrough (SWB) (Scheck and Ross 2008).

A plot of PCA1 vs. time indicated a change occurring at a specific point in time, which was attributed to seawater breakthrough. A comparison with theoretical values for seawater/formation-water mixtures revealed seawater breakthrough of 8–10%.

Another use of produced-water compositional analysis is in cases where nitrate injection is used for reservoir-souring control. In these cases, the aim of the nitrate injection is to allow all the carbon food source to be consumed by nitrate-reducing bacteria so that none is left for consumption by SRB. As a consequence, no sulfate is converted to sulfide, thereby preventing reservoir souring (Maxwell 2007).

One of the difficulties associated with this treatment philosophy is the uncertainty regarding the amount of carbon sources available for consumption by bacteria. Volatile fatty acids have been traditionally assumed to be the primary SRB food source, but it is also possible that aromatics such as benzene and toluene might be used. These molecules have some solubility in both oil and water. Therefore, their consumption from the water phase by bacteria could provide an extremely large source of carbon because it would presumably result in the partitioning of more of those molecules from the oil into the water such that a solubility equilibrium is maintained.

As a result of the impact of benzene and toluene partitioning, there could be considerable uncertainty regarding the nitrate concentrations needed for full souring control. The best way to manage this issue is to perform surveillance in which the produced water is analyzed for nitrate ion (which is typically absent in formation waters). A large excess of nitrate after injection-water breakthrough would allow a reduction in the chemical-injection rates. Its absence would suggest that nitrate injection needs to be increased if full souring control is to be achieved.

In addition to compositional data, surveillance of scale squeeze treatments invariably also incorporates analysis of scale inhibitor returns to identify when retreatment is needed. Such analyses were required for the North Sea Miller Field, where different squeeze chemistries were being evaluated (Bourne et al. 2000). **Fig. 7** shows the inhibitor returns for two different chemistries, from which it can be observed that the new inhibitor chemistry delivered a 3.5-fold increase in squeeze life.

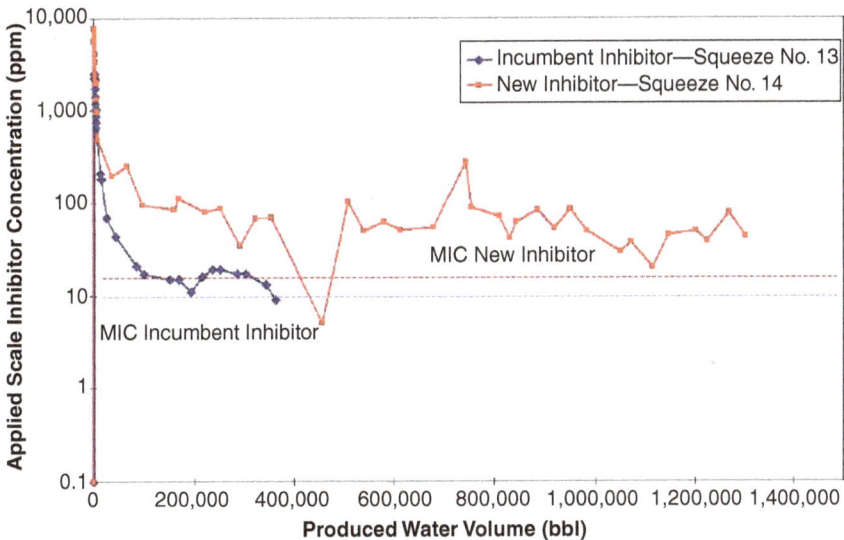

Fig. 7—Scale inhibitor return measurement for squeeze evaluation (Bourne et al. 2000). MIC = minimum inhibitor concentration.

Historically, many assets have struggled with the difficulties in measuring inhibitor returns. Those difficulties are magnified where squeeze treatments are performed in subsea wells because the measurements must be performed on commingled fluids from many wells. As a result, the concentrations are inevitably lower than they would have been in a single-well stream. In such cases, the inhibitor residuals quantification might become even more important. This is because the measurement of scaling ions can no longer be used for diagnosis because of the presence of fluids from other wells.

Ion chromatography mass spectroscopy has been used to deliver improved sensitivity quantification of phosphonate scale inhibitor returns in this setting (Zhang et al. 2014). Another approach is to chemically modify the scale inhibitor itself so the backbone of the inhibitor polymer can be detected by fluorescence spectroscopy (Vuori et al. 2016). This technology offers the prospect that inhibitor residuals could be detected on-site through the use of a spectrofluorometer.

3.2.5 Corrosion Monitoring. Corrosion-rate monitoring in both the injection and production systems is needed to ensure that the process equipment, flowlines, and wells have lifetimes in accordance with design expectations. Corrosion processes usually occur relatively slowly, so the response time is usually much slower than it is in other systems. The monitoring program safeguards against premature system degradation that could either compromise longer-term production or induce extra costs for replacement of the equipment. However, there could be cases in which a much quicker response to observed problems is required. This could occur when matrix injection is a requirement and poor water quality induces increased corrosion, which causes increased levels of suspended solids. A faster response could also be required when hydrogen sulfide (H_2S) levels have the potential to induce stress cracking problems because this corrosion type can induce failure far more quickly than conventional corrosion processes.

The design and setup of the corrosion monitoring system need to consider the methodologies, location, and required frequency of corrosion surveillance. These requirements could differ at different locations in the process on the basis of anticipated risks at each location and the type of corrosion expected. Although the corrosion engineer is likely to take the lead in defining that program, a number of other resources should also be involved, including production chemistry, pipeline engineering, facilities engineering, and production operations.

From a system integrity perspective, many corrosion monitoring programs adopt a risk-based approach such that system surveys are scheduled on the basis of anticipated corrosion rates. Improved monitoring accuracy might reduce uncertainty in the expected corrosion rates, enabling a less conservative estimate of corrosion rates to be used in defining the inspection frequency. However, this approach might not be appropriate in cases where the water quality of the injection system is a critical surveillance element.

The corrosion monitoring program must define the tools used to measure corrosion rates. One option is to measure the physical metal loss, either on a sensing element or on the metal surface itself. Methodologies that fall into this category include weight-loss coupons, ER probes, ultrasonic thickness measurements, and pulsed eddy current measurement. Weight-loss coupons are probably the most commonly used surveillance method because they are inexpensive, offer a reliable guide to corrosion rates, and can be used under steady-state corrosion conditions where electrical devices cannot be used.

Another surveillance method is to measure the electrochemical properties of the corroding surface. These techniques include LPR probes, galvanic probes, electrochemical noise probes, potential measurement probes, and potentiodynamic polarization probes. They require that there be a wetted electrode surface area for determination of the current density and a sufficiently conductive area between the electrodes. Sometimes it is difficult to meet this requirement in the system itself, and measurements are conducted in a sidestream.

LPR probes give a continuous indication of corrosion rates in electrically conductive fluids, making them useful corrosion monitoring probes in water-injection systems. However, they cannot be used in an environment where oil is present, so they are not appropriate for use in production systems. The probes might protrude into the water line, or they might be flush with the pipe wall. Galvanic probes are commonly used in water-injection systems as indirect oxygen monitors. Although they do not actually measure oxygen, the bimetallic couple responds rapidly to changes in dissolved oxygen levels. Thus, they are often used in a high-pressure system within a water-injection system.

Another group of techniques measures the byproducts of corrosion. This includes hydrogen probes, which measure atomic hydrogen from the cathodic corrosion reaction. This technique is usually deployed when hydrogen-induced cracking is considered a risk. Another technique that falls in this area is to measure the amounts of dissolved and total iron. This is usually used as a supplementary, rather than the primary, method of surveillance because long-term data trends are needed, and it only gives an indication of the average metal loss over the whole system unless the source of iron can be pinpointed. It can only be used to detect variations in corrosion rates, not absolute corrosion rates. Therefore, it is typically used as a backup monitoring method.

When used in the production system, the measurement of iron concentrations will have limited value if the formation water contains significant levels of dissolved iron or when the maximum iron concentration is limited by the solubility of iron salts (oxides, sulfide, or carbonate) rather than by corrosion, resulting in low readings. Iron-count monitoring is not recommended in sour systems because sulfides could precipitate, giving a lower dissolved-iron concentration than is representative of the true corrosion rate.

Visual inspection can be a very important part of the corrosion monitoring program, providing useful information on both the extent and type of corrosion. Any time that well tubing is retrieved, the opportunity should be taken to identify the type and extent of corrosion. Spool pieces (short, flanged pipe sections) can also be used and installed with block-and-bypass valves so that the internal condition (e.g., pitting) can be correlated with other corrosion monitoring data.

In selecting the monitoring techniques to be used, it is important to consider which techniques are suitable for monitoring at the prevailing conditions of temperature, pressure, and fluid composition. The type of system access that is needed (e.g., nozzle, spool, sample point) also should be considered. It is then important to consider whether the corrosion is expected to be uniform or whether there might be more localized corrosion hot spots. How the planned monitoring locations relate to the generalized corrosion rates and the rates at the anticipated locations of the most severe corrosion is also important. Corrosion caused by CO_2 or H_2S generally tends to be localized in pits or grooves, and the observed corrosion rates tend to be more severe in areas of high turbulence, such as at bends or valve locations. In addition, the region around welds is often more susceptible to attack than the parent metal.

The corrosion monitoring of wells can also be an important element of the surveillance plan because it is important to verify that wells will be available in the longer term to secure long-range forecasts. Corrosion logging can therefore be an important element of a surveillance plan (Jalan et al. 2013). It enables an assessment of the conditions of both casing and tubulars, but given the costs associated with such surveys, they are likely to be performed by exception rather than as part of a standard surveillance program. Monitoring the annular pressures in a well is likely to be a standard part of a surveillance program to verify that there is no direct tubing-to-annulus communication.

In some cases, corrosion problems can be initiated when mature waterfloods begin to use electrical submersible pumps (ESPs) to improve lift performance as water cuts increase. This problem was experienced in a number of wells in Oman as a result of stray current leakage from the ESP power cable or when one of the phases in the ESP motor was not functioning, creating a phase imbalance. The stray currents provide the driving force for metal loss on motor housings and well casings (Al Mahrooqi and Azim 2015). The installation of zinc anodes below the ESP motor has been effective in mitigating motor-housing corrosion. When casing corrosion problems manifest, it might be necessary to install a cathodic protection system.

Ultrasonic and electromagnetic logging was instrumental in identifying the nature of the problems in this case. **Fig. 8** shows an example of corrosion logging

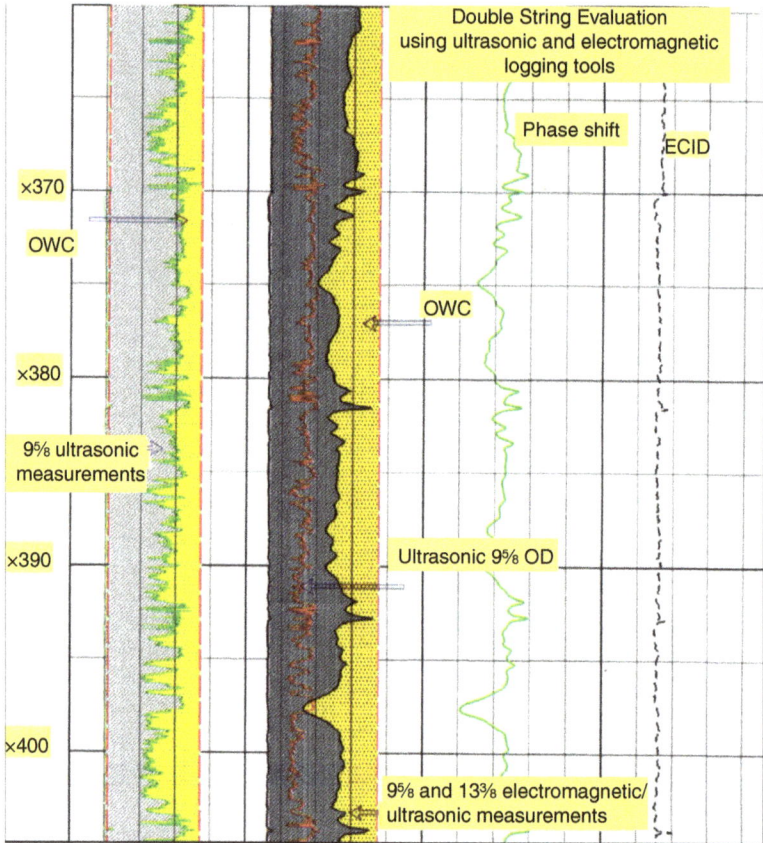

Fig. 8—Corrosion logging example (after Al Mahrooqi and Azim 2015). OD = outer diameter; OWC = oil/water contact.

where the production casing suffered from external circumferential damage (second track from the left). Electromagnetic measurements for combined casings indicated a complete destruction of the 13⅜-in. casing (third track from the left). This example demonstrates that well corrosion logging can sometimes be a vital element of a surveillance plan to understand the nature and location of well corrosion problems.

3.3 Well Surveillance.
3.3.1 Injectors.
Injectivity Trends. Monitoring injection rates and pressures is a fundamental part of a standard waterflood surveillance program. A plot of the pressure and rate history of an injector over an extended period can help diagnose injectivity problems. Measurements of bottomhole pressures are usually preferred because measurement at the downhole location removes any uncertainty associated with frictional pressure losses at different injection rates. These will be readily available if downhole gauges are installed. Surface pressures can also still give some indications of well behavior, but it would be preferable if the engineer were able to take account of the frictional pressure loss.

A very useful data display is the Hall plot. It shows the cumulative pressure time product against cumulative injection (Jarrell and Stein 1991). This plot is useful to detect changes in an injection well's flow capacity as indicated by a change in the slope of the plot (**Fig. 9**). A straight line on the Hall plot normally indicates constant injectivity. A decrease in slope is an indication of increased flow capacity as a result of opening a new zone, fracturing, or decreased skin. An increase in slope is an indication of plugging or other wellbore damage. The relative amount of improvement or damage can be estimated by comparing the change in slope.

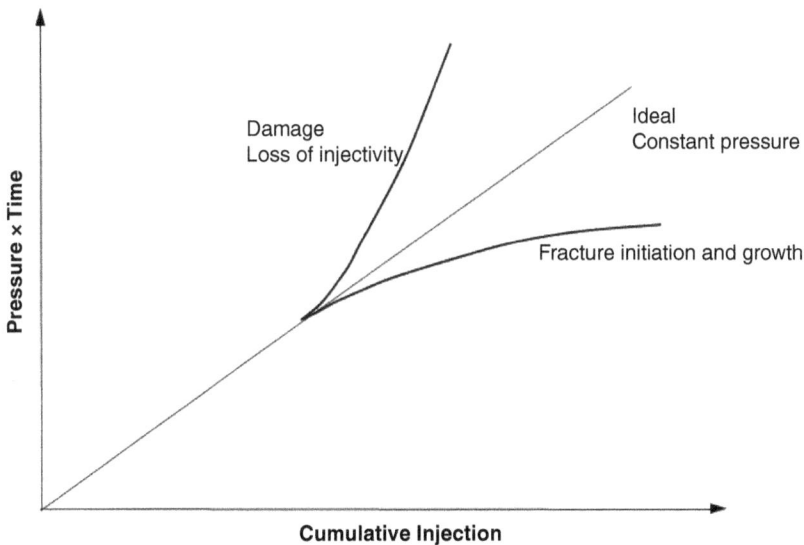

Fig. 9—Hall plot.

Hall plots must be interpreted carefully because a trend similar to that seen when a well begins to fracture will also be observed if there is a gradual decline in the reservoir pressure (Amedu and Nwokolo 2013). Often, a Hall plot uses the summation of the recorded injection pressure, which leads to this problem. A more rigorous Hall plot should take account of the reservoir pressure. In that case, the Hall integral (I_H) is given by

$$I_H = \int (P_{wf} - P_e)\, dt, \quad\dots\dots\dots\dots\dots\dots\dots\dots\dots\dots\dots\dots\dots\dots\dots\dots\dots\dots \text{(1)}$$

where P_{wf} and P_e are the flowing bottomhole injection pressure and reservoir pressure, respectively.

Similarly, relative permeability changes that occur as the flood progresses can also induce changes to the slope of the Hall plot. For injectors that are completed in the oil column, a gradual improvement in injectivity might be expected over time because of relative permeability changes that increase water injectivity (Wu and Ershaghi 2019). Such effects can be particularly significant in viscous oil systems. Therefore, data must be interpreted carefully because changes that improve injectivity are normally regarded as being indicative of fractured injection.

Another issue that might induce problems in interpretation can occur if the benefits of injectivity changes resulting from relative permeability are balanced by an increase in damage caused by suspended material. In this event, a straight-line Hall plot would be maintained, giving an erroneous indication that nothing is changing.

Sometimes, there can also be problems in the interpretation of Hall plots in layered reservoirs (Qi et al. 2017).

One of the problems with Hall plots is that changes in slope are often somewhat subtle, and it is not always easy to discern changes in the flow regime. Those limitations have been removed somewhat through the use of the Hall plot derivative (D_{HI}) (Izgec and Kabir 2007):

$$D_{HI} = \frac{d\int (P_{wf} - P_e)\, dt}{d\ln(W_i)}, \quad\dots\dots\dots\dots\dots\dots\dots\dots\dots\dots\dots\dots\dots\dots\dots\dots\dots \text{(2)}$$

where W_i is the cumulative water-injection volume.

In this derivative plot, upward changes in derivative signify a loss of injectivity. Downward changes in derivative signify injectivity improvement. The derivative is normally plotted alongside the traditional Hall plot for comparison purposes. **Fig. 10** shows an example from an injector in the Akpo Field (Onwuchekwa et al. 2019).

If the derivative and the conventional integral are on top of one another, it is an indication that nothing is changing, and matrix injection is assumed. When the derivative is below the conventional Hall plot, it indicates fractured injection or channeling. It has been suggested that a constant separation distance between the derivative and the integral indicates channeling. The derivative above the conventional plot is an indication that plugging is occurring.

The Hearn plot is another method that uses the reciprocal of the injectivity index to assess the injection performance of a well. **Fig. 11** shows the corresponding Hearn plot for the same well as in Fig. 10.

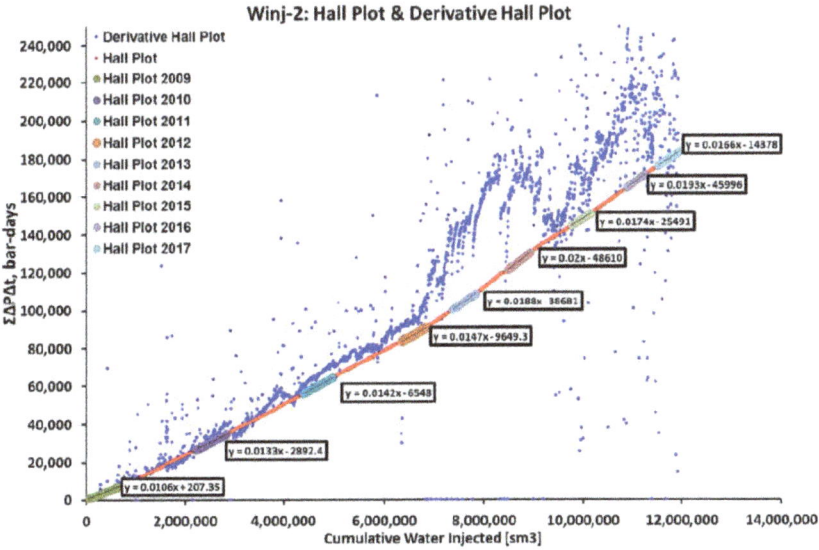

Fig. 10—Hall plot derivative (Onwuchekwa et al. 2019).

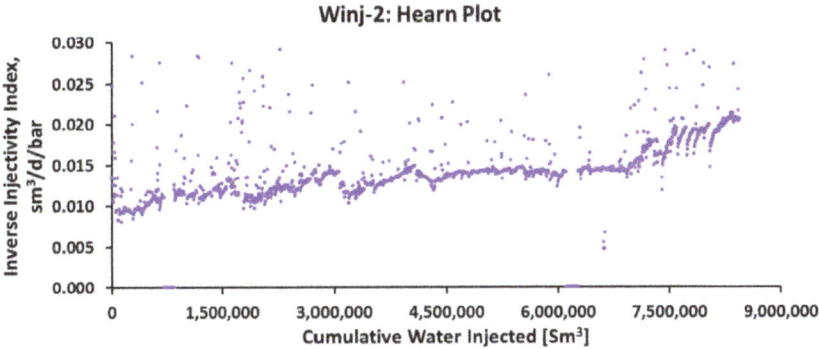

Fig. 11—Hearn plot (Onwuchekwa et al. 2019).

Injection Metering. Injection metering is fairly straightforward because water injection is a single-phase environment, which removes one key area of uncertainty. Generally, therefore, it is usually reasonable to expect a high degree of accuracy in such measurements. Nevertheless, it is appropriate to perform an injection reconciliation between the individual injector-volume measurements and a bulk-volume meter. This reconciliation can sometimes indicate a problem with a measurement that requires attention so that the relative proportions of the injected water volumes taken by each well can be accurately known. Additionally, stability in the tracking of the reconciliation factor is important because changes can indicate an element is no longer functioning correctly.

Pressure Transient Analysis. Falloff testing of injection wells is a vitally important part of the surveillance plan, particularly when injection takes place under fractured-injection conditions. Information can be derived regarding the induced fracture dimensions and how they might influence the flow pathways through the reservoir. This topic is discussed in detail in *Waterflooding: Injection Regime and Injection Wells*, another book in this series, which discusses the operation of the injection wells.

Injection Logging. Injection production logging tool (PLT) logging can be a valuable technique for understanding injection distribution. Because temperature is one of the elements reported in a standard PLT logging suite, these data can be critically important when collected after a well has been shut in as the wellbore warms back toward the geothermal gradient. This warmback technique can identify any out-of-zone injection. (See *Waterflooding: Injection Regime and Injection Wells*, another book in this series, for additional discussion.)

Injection PLT identifies where the water leaves the injection wellbore, but this might not always be the location where the water enters the reservoir because there is sometimes the possibility of annular flow. It is possible to identify such problems using noise logs in combination with temperature logs (Galiev et al. 2018). Sensitive, passive hydrophone tools run through the target injection zone are able to identify where the injection water travels, and any behind-pipe flow can usually be identified in this manner. The full quantification of the direction of flow in complex cases is aided by the use of temperature logs because there can be a Joule-Thomson cooling effect, and logging at different injection rates will further improve the diagnostic capability.

Fig. 12 illustrates one such logging exercise where noise logging, temperature logging, and conventional PLT data were used. It shows that 77% of the injected water

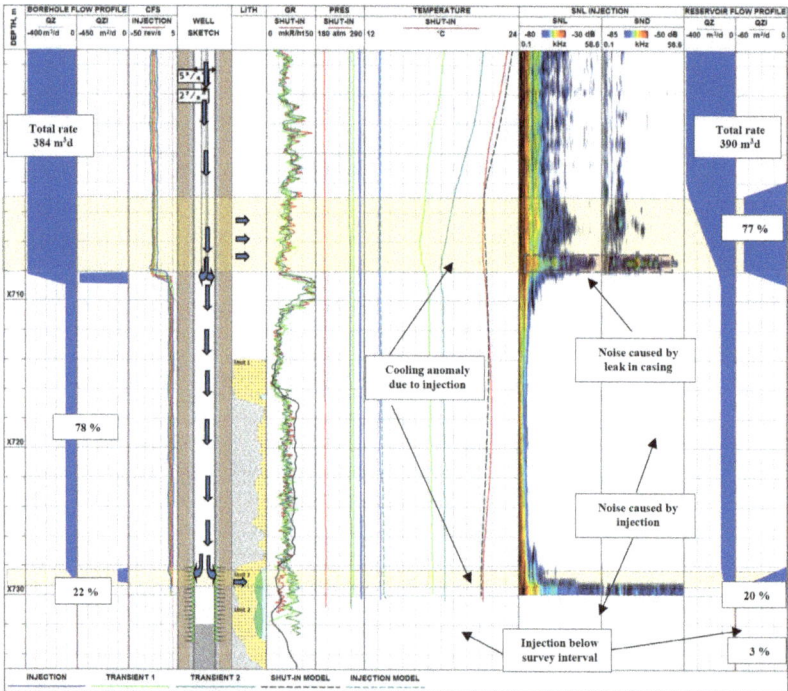

Fig. 12—Noise logging showing injection through a casing leak above the perforated interval (Galiev et al. 2018). SNL = spectral noise logging.

was entering an unperforated interval through a leak in the 9⅝-in. casing. According to the PLT data, significant injection loss was occurring across the wireline entry guide, suggesting either casing or packer integrity issues. Subsequently, a cement squeeze was performed. This resulted in an overall reduction in the injection rate, but it was evident that the problem was resolved because the offset production rate increased by more than 10%.

Another example is shown in **Fig. 13**. In this case, the logging of a horizontal injector demonstrated that 91% of the injected water went into the top of the slotted liner and only 9% entered the remaining intervals. The analysis showed that out-of-zone injection was occurring by means of flow behind the casing, and the nature of the flow suggested the likely presence of an open-fracture system. The well injectivity was subsequently found to be significantly lower than expected. This was possibly the result of reservoir-quality degradation with depth because the next well was drilled shallower and performed as per expectations.

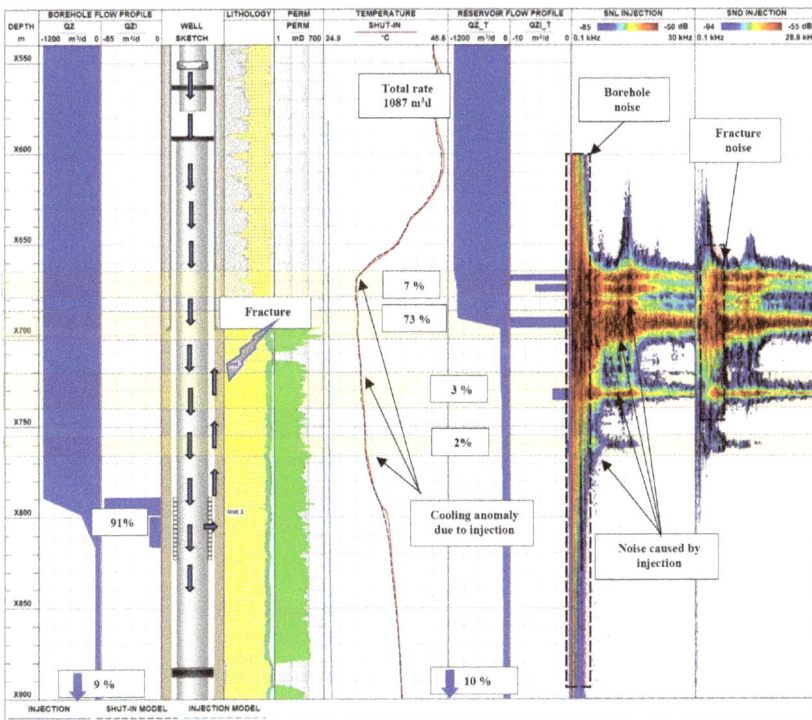

Fig. 13—Out-of-zone injection through fracture (Galiev et al. 2018). SNL = spectral noise logging.

PLT data can be difficult, and expensive, to obtain in horizontal injection wells. In some cases, this can make the installation of fiber-optic cable an appropriate alternative. This option is being actively pursued for the surveillance of injection wells in California's Belridge Field (Allan et al. 2013). The fiber is deployed in new wells outside the casing and cemented in place, but in some existing wells, fiber use was also possible by means of running the fiber inside the tubing. The primary challenge has been in the interpretation of the data output. In this case, the operator built its own software to interpret the signals, reducing the cost of data acquisition by 95%.

With this technology, a light pulse is emitted by a laser at the wellhead and then moves along the fiber. Fragments of the light strike the fiber wall and are reflected back to the data acquisition at the surface, where the Raman backscatter band is used to measure temperature. This technique can be used in three different ways to evaluate injector performance (Rahman et al. 2011):

- Stabilized injection: Temperature profiles are recorded with the well injecting at a steady rate. For this method to be successful in detecting injection profiles, the injection-water temperature must be materially different from the geothermal gradient across the injection interval. The viability of the technique can also be affected by the permeability distribution in vertical wells. Where the uppermost interval takes the bulk of the water, the distribution below can still be seen, but when the zone taking most of the water is deep within the interval, it can be difficult to identify the injection distribution to the zones above.

- Thermal restoration (warmback): This technique requires the injector to be shut in long enough to see which zones return to the geothermal gradient. Zones that have taken more water will remain cooler against the geothermal gradient. This can be used as a qualitative indicator of the injection profile and can be very useful as a guide to out-of-zone injection.

- Thermal tracer: This technique generates a temperature signature using injection water that has a different temperature than that of the normal injection water. Analysis is easier where this temperature difference is greater. A hot or cold injection slug is added, and then, as regular injection resumes, the fiber can track the hot/cold interface as it moves down the well and exits through the perforations.

At Belridge, this latter technique was identified as the most appropriate for waterflood surveillance. It was used to monitor subsurface injection conformance, changes of profile with time, and the optimization of the waterflood. This was achieved through the identification of wells requiring the correction of thief zones and by diverting target injection volumes preferentially to wells with superior injection profiles.

It might also be possible to use distributed temperature sensing to monitor injection profiles in horizontal injection wells (Bui and Jalali 2004). Shut-in temperature data can be used to estimate the injection profile on the basis of a history-matching technique that relies on existing analytical solutions for vertical wells. After shut-in, the data can be used to estimate the ongoing injection profile, although this could have a low spatial resolution.

3.3.2 Producers.

Water Composition. The compositional analysis of produced water can be a very important parameter to aid the identification of injection-water breakthrough. It is then a very useful parameter for the history matching of the reservoir model and serves as a guide to estimate whether the reservoir sweep is as good as was predicted at the outset.

For this methodology to be viable, there should be some compositional difference between the injection water and the formation water. It can therefore be unsuitable for cases in which only produced water is used for injection. In those circumstances, other techniques such as tracer testing will be needed to quantify reservoir transit times. When there is an appreciable difference between injection water and formation water, the first step is to look at the two compositions to ascertain which ions have sufficiently different concentrations to qualify as markers. Some ions could become depleted because of scaling reactions in the reservoir, so it is best not to rely on the measurement of only one ion to determine the exact breakthrough timing. For a seawater-injection scheme, chloride, potassium, and magnesium are often good ions to use as markers.

This was performed at the North Sea Miller Field, where the chloride ion concentration was used as a tracer to detect seawater-breakthrough timing (Houston et al. 2006) (**Fig. 14**). The water-breakthrough timings in this field showed appreciable differences, suggesting that the permeability in the reservoir was not homogeneous.

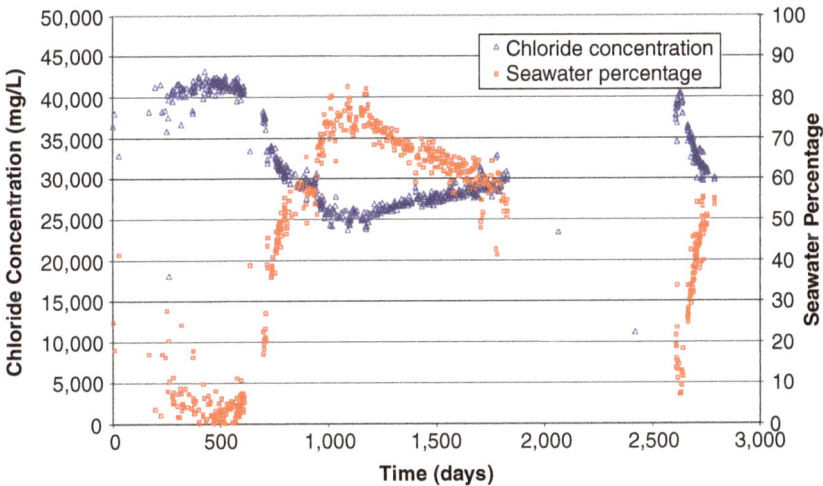

Fig. 14—Changes in chloride ion concentration to detect seawater breakthrough (Houston et al. 2006).

Because compositional analysis was already being performed, it was possible to look at the changes of other elements in the produced water. This showed that calcium was enriched and that magnesium, barium, and sulfate were each depleted relative to the normal mixing ion concentration expected based on the relative proportions of seawater and formation water (**Fig. 15**). These data showed that calcite dissolution and replacement by dolomite and barium precipitation were occurring within the reservoir.

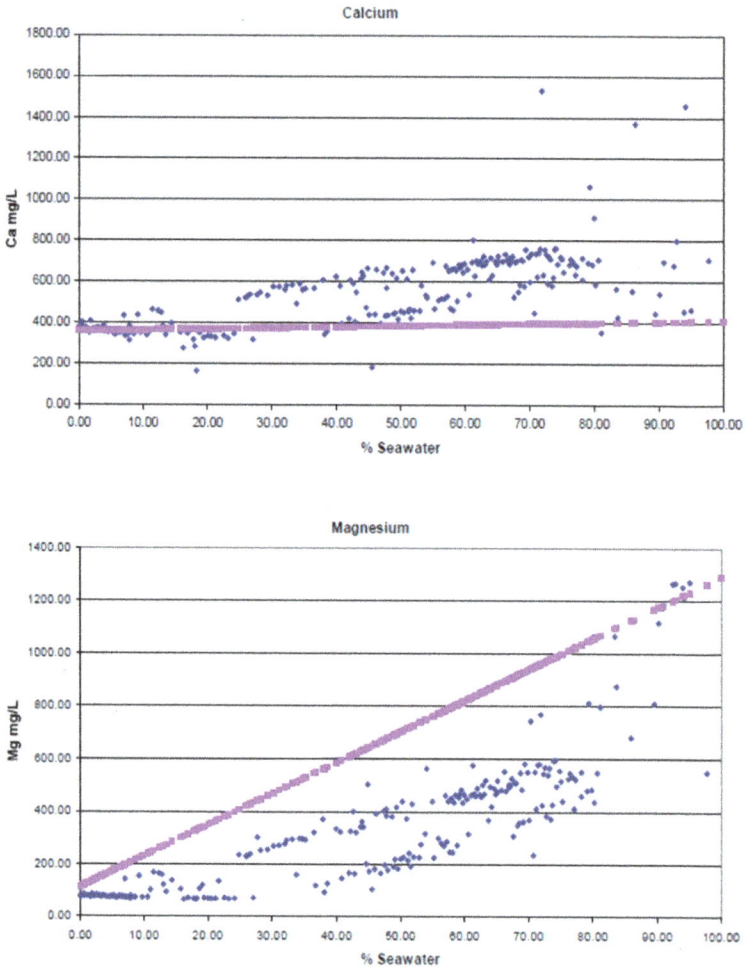

Fig. 15—Calcium (Ca) and magnesium (Mg) ion concentrations relative to a mixing line (Houston et al. 2006).

Reservoir-Souring Monitoring. Reservoir-souring monitoring is usually based on an analysis of samples in the gas phase. Early detection of reservoir-souring onset is important if there is no reservoir-souring protection program in place because this could facilitate the commencement of an appropriate remediation program, in addition to an H_2S scavenging program. Initially, H_2S concentrations will be low, so they might not be easily detected if surveillance is based on bulk-fluid measurements. It is thus best to base such a program on individual-well measurements. It might be impractical to perform measurements in all wells, so it is appropriate to design a program based around more frequent measurement in the higher-risk wells. Based on the reservoir-souring mechanisms, it is most likely that H_2S would first be found in wells that experience water breakthrough early and/or wells in which there is a high pore-volume throughput of water.

The simplest method for measurement of H_2S in gas uses colorimetric sample vials similar to those used for oxygen measurement. In this case, the tube responds to H_2S by progressively darkening. Different tubes can be purchased to reflect the anticipated concentration range, so low-range tubes should be used if the field has not yet soured.

These analyses have the advantage of being simple, so they can easily be accommodated. However, they do have limitations in terms of accuracy (±20%), and changes in temperature, in addition to interference from other chemicals in the process, might influence the measurements. Consequently, performing periodic cross checks with more-accurate gas chromatographic techniques might be beneficial. Samples for that analysis will probably need to be sent to the laboratory in sample bombs. In that case, polytetrafluoroethylene-lined sample bombs must be used to avoid potential H_2S adsorption within the sample vessel, affecting the subsequent results.

Because H_2S will partition between the different phases, the H_2S present in the produced-water phase should also be measured. Accurate volumetric analysis then allows for the quantification of the partitioning coefficients.

Production Profiling. Shutting off unwanted water production is often a very important aspect of optimization in waterflooded fields. It is therefore important to identify water-influx locations in wells that produce water. In vertical or slightly deviated wells, this is usually achieved through standard PLTs, which can be run in hole on wireline or coiled tubing and can be tailored to either memory or surface readout configurations. The tools that are run can be tailored to the specific requirements of each application, but a standard PLT logging suite would typically incorporate the following:

- Flowmeter: The spinner measures flow rate, with the revolutions per second being proportional to the flow rate. Metering can be either continuous, using full-bore flowmeters, or stationary, with point profiling by petal basket or packer flowmeters. The lower limit of flow rate that can be reasonably measured with full-bore spinners is normally in the range of 250 to 500 B/D, depending on the hole size. The stationary systems use a convergent element that forces flow into the metering chamber, where it turns an internally mounted spinning device. Fluid rates as low as 100 B/D can be measured with reasonable accuracy using this methodology. It becomes less accurate at high flow rates as a result of fluid bypassing the tool. Furthermore, the tool must be stationary before a reading can be taken.
- Gradiomanometer: A gradiomanometer measures density, on the principle that the difference in pressure between two points in a column of fluid is a function of the height of the column and the fluid density. Two sensors at different heights measure the pressure differential, and from this the fluid density is calculated. Because the tool relies on the hydrostatic gradient, it cannot be used in wells with deviation angles greater than 60°. For such applications, nuclear fluid densimeters can be used through the measurement of the absorption coefficient of gamma rays, which is proportional to density. Gamma rays are emitted from a source, and the absorption spectrum is measured by a detector located a short distance away from that source. Data acquisition with this tool requires a slow logging speed.
- Pressure/temperature sonde: Pressure and temperature are usually measured simultaneously using this tool. Gradients for both parameters over the length (or part of the length) of the tool are also measured.
- Caliper: A caliper measures the internal radius of the borehole. This measurement is needed to convert fluid velocity to a volumetric flow rate.

- Gamma ray log or casing collar locator: This is needed to provide depth correlation. The casing collar locator measures the change in magnetic flux caused by variations in the casing thickness at joints. The gamma ray log measures natural radiation emitted by the lithology; when used as part of a PLT configuration, it is run through the casing and correlated against the gamma ray run in the open hole.

PLT interpretation in low-deviation wells is fairly straightforward, but in very high-deviation and horizontal wells, the standard logging suite is unable to deconvolve the relative flow influx into the wellbore. This is because phase segregation induces fluid holdup in the wellbore. In horizontal wells, the flow stratifies because of fluid density differences, and this materially influences how the fluids flow with changes in deviation, especially at low flow rates.

It follows that tools that are able to deconvolve the phenomenon of holdup need to be able to measure the phase regime at all points in the pipe, in addition to the flow velocities. These tools are available from service companies, although they are more expensive to run than the standard PLTs used in vertical wells. Furthermore, while wireline can run the tools in vertical wells, the deployment of logging tools in horizontal wells requires either coiled tubing or tractor deployment, adding further cost (Oosthuizen et al. 2007).

Because of the increased cost of logging in horizontal wells, it is likely that data collection is less frequent than in vertical wells. However, the advent of fiber-optic technology raises the prospect of constant productivity monitoring. This would facilitate just-in-time intervention and eliminate the need for the costly re-entry and logging processes. However, although there has been progress in the interpretation of fiber-optic data relating to two- and three-phase fluid flow (Kortukov et al. 2019), it is not yet evident that such data can unambiguously quantify water-influx points with the same accuracy currently provided by PLTs.

Pulsed Neutron Logging. Pulsed neutron logging can be used to identify cement channels in producers when they are responsible for water short circuiting. First, a base pulsed neutron log (PNL) pass is conducted before injecting a borax solution into the well. Then, additional PNL passes are conducted.

The PNL tool emits pulses of high-energy neutrons into the formation. Through interaction with the nuclei of different atoms present in the borehole and formation, they lose their energy. This process is accompanied by a release of gamma rays by the logging tool detectors. Of the two boron isotopes present, ^{10}B has a capture cross section that is a few orders of magnitude greater than that of most of the naturally occurring elements in the formation. Therefore, any part of the formation invaded by a borax solution will show a higher capture cross section compared to the base conditions.

If there is no cement channel, the injected borax solution only flows laterally through the formation, with no vertical component. However, if a cement channel is present, the borax solution will move upward or downward before being absorbed into the formation, resulting in a higher measured capture cross section during the borax pass than that of the base-pass value. By comparing the post-borax signal with the base pass using certain assumptions, any increase in formation response is attributed to borax. This is inferred to be caused by the flow of borax channels behind the casing.

There are numerous examples of successful deployments of this technique (Blount et al. 1991; Barnette et al. 1992; Sommer and Jenkins 1993). The interpretation of such logs assumes that any vertical flow of the borax solution is caused by cement channels only. While this might be true for reservoirs with interbedding shales, it might not be valid for other reservoirs where shale barriers are absent. Das et al. (1998) reported that in an application in a carbonate reservoir the technique was less definitive, possibly resulting from karstification in the reservoir. Even so, with a detailed understanding of the reservoir geology, some interpretation of the borax signals might still be possible.

Well Testing. Production testing provides data fundamental to interpreting the efficiency of the flood's progress. The testing frequency must be high enough to allow a reasonable quantification of when water breakthrough occurs because this will provide the earliest indication of the flood efficiency. It also provides an important calibration point for reservoir models. Ideally, all producers should be tested on a monthly basis. Frequent, high quality, well testing can also identify cases where there is a stepwise development in increasing water cut. This can indicate water breakthrough in different layers. Unfortunately, sometimes problems with separators in mature fields can limit the frequency at which wells are tested—just at the time that well testing assumes critical importance. In such cases, the testing frequency for each well might need to be prioritized, with the most important wells being tested the most frequently.

A range of well testing options is available, including an automated three-phase test separator, a multiphase flowmeter, a Coriolis meter, and a test tank (onshore). Improved methods might be needed to measure the net oil produced as the water cut reaches high levels. This is because many methodologies carry significant inaccuracy in the oil cut measurement when high water cuts are reached. This can be important because many waterflood wells are abandoned on the basis of what is deemed to be a water cut at which the well becomes subeconomic.

Decline Analysis. Decline curve analysis to extrapolate production well perfor-mance can be used in mature waterflood wells to assess remaining reserves. Such analysis must be used carefully, especially if the flood is relatively immature. This is because events such as fill-up or oil banking will significantly affect oil production trends and are thus likely to render the extrapolation of current performance as a guide to future performance ineffective.

In more mature waterfloods where the water cut is reasonably developed, vol-umetric sweep is likely to be much more stable and decline analysis is much more likely to be a reliable technique. At this time, the changes in oil production rates will be primarily driven by relative permeability effects.

The criteria for the application of decline analysis to waterflood producers there-fore include

- Reasonably significant water-cut development.
- Stable gas/oil ratio (GOR) production implying adequate and stable voidage replacement performance.
- Relatively constant injection and production rates.
- Absence of infill: If a waterflood pattern is infilled, the volumetric sweep will be affected, so decline analysis will be affected for some time.

In many waterfloods, a plot of log[water/oil ratio (WOR)] vs. cumulative oil tends to give a straight line provided that production, injection, and pressure are stable, the producing water cut is greater than 50%, and there is full voidage replacement.

Analysis of waterflood producer declines suggests that harmonic or hyperbolic declines are likely to be appropriate. However, exponential or superexponential declines are also possible (injectors impairing). Superhyperbolic decline is also possible in layered reservoirs, although some wells could exhibit exponential declines.

Sometimes, decline analysis can provide useful pointers on waterflood performance. For example, a difference between the oil rate and oil-cut recovery extrapolations could indicate underinjection.

3.4 Reservoir Surveillance.

3.4.1 Pressure Data. High-quality pressure data are fundamental to understanding waterflood performance because it is pressure that drives fluid flow. Flux occurs from areas of high pressure to areas of lower pressure. While there will clearly be a pressure differential between injectors and producers, it is preferable to otherwise have relatively uniform pressure deep within the reservoir.

A range of different types of pressure data can be obtained as part of a waterflood.

Repeat formation test measurements are typically taken when infill wells are drilled, to update the understanding of the pressure distribution within the reservoir. Such data can be particularly useful in assessing the degree of vertical sweep efficiency in stacked or multilayered reservoirs.

Pressure buildup and static pressure surveys are normally built into the routine monitoring requirements for a field. Such data could be important to assess if any areas of the waterflood are not receiving adequate pressure support, and they are readily collected at minimal cost. Flowing pressure surveys can be run successfully if the well is producing under reasonably stable conditions. They are typically conducted at several locations within the well. Pressure buildup surveys involve a well that is already flowing (ideally at a constant rate) being shut in, then observing downhole pressure trends as pressure builds. This provides information on reservoir pressure and the well productivity index.

Downhole gauges are often installed in deepwater wells. Therefore, a pressure buildup trend (or, alternatively, a pressure falloff trend for an injection well) can be obtained every time the well is shut in.

Pressure surveillance data can be vitally important in injection wells in particular. In a matrix-injection scheme, the injection pressure must be kept low enough to avoid fracturing the well. In a fractured-injection scheme, the pressure must be maintained below the fracture pressure of the caprock to ensure reservoir containment.

3.4.2 Saturation Measurement. Techniques for saturation measurement in wells can be used in injectors, producers, or even observation wells. Time-lapse logging of observation wells can reveal the progression of saturation fronts in various layers. It can also show the upward migration of water in gravity-dominated situations. Such measurements can be particularly important in oil-wet reservoirs where significant pore-volume throughputs could be required to approach residual oil saturation.

Traditionally, logging for saturation measurement has used either carbon-oxygen (CO) or thermal decay time (TDT) logging methods. TDT logging looks at the rate of thermal neutron absorption. A high absorption rate indicates saline water containing chlorine ions, which are very efficient thermal neutron absorbers. A low absorption rate indicates either fresh water or hydrocarbons. This technique therefore cannot be effectively used in low formation-water salinities, where it is difficult to differentiate between oil and water. In such cases, CO logging could be used. This technique measures gamma rays emitted from inelastic neutron scattering to assess the relative proportions of carbon and oxygen in the formation. A high CO ratio indicates oil whereas a low CO ratio indicates water (or gas). The problems with this technique are its sensitivity to borehole fluid and the fact that a large borehole size might be needed to accommodate the tool.

These limitations have been largely overcome with a new generation of tools offering improved signal-generation and -detection capabilities. A number of such tools have been compared for application in a Middle Eastern carbonate reservoir (Eyvazzadeh et al. 2004).

Field measurements can be useful to reconcile log data with laboratory-based measurements of residual oil saturation. Such a study for the North Sea Cormorant Field found that core-based measurements can be improved by first cleaning the cores, then aging samples in the presence of oil and water for a number of weeks before measurement (van Poelgeest et al. 1991).

Such logging runs in producing wells could provide information regarding the distribution of the flood profile. They also provide information on the amount of the in-place oil that could be swept on contact with injected water.

Behind-casing resistivity measurements in producers have been used in a mature waterflood at the Elk Hills Field in California (Starcher et al. 2002). In this case, the measurements helped to identify unswept oil intervals in a number of supposedly watered-out wells. Consequently, following isolation and reperforation, the wells in question were able to be successfully put back on production.

3.4.3 4D Seismic. A 4D seismic survey entails performing repeats of a baseline 3D survey to enable the quantification of the effects that production and injection have on the seismic profile. Consequently, 4D seismic monitors the changes in 3D seismic surveys as a function of the fourth dimension, time. As the fluids move within the reservoir, the fluid type at any given location will change. This will induce a change in the seismic response (**Fig. 16**), and pressure changes as a result of production and injection could also influence the seismic response. In waterfloods, the replacement

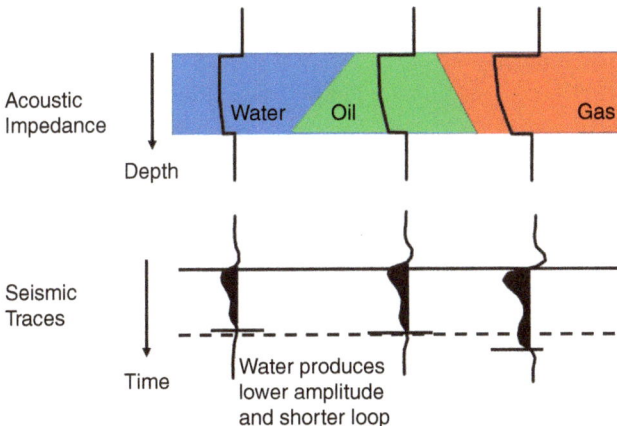

Fig. 16—Effect of fluid saturation change on seismic response.

of oil by water as a result of injection is termed hardening, and it induces an increase in acoustic impedance. Such changes can also be induced by a drop in the reservoir pressure.

The technique thus offers a means to visualize hydrocarbon saturations and pressure changes across the field as a result of waterflood movements. Consequently, it provides a very powerful tool that can assist in ongoing waterflood optimization.

Seismic resolution, of approximately 10–20 m, is small with respect to most well spacings. Therefore, 4D is a very useful indicator of infill opportunities and the lateral movement of water in the reservoir. However, the resolution makes it less useful for understanding the vertical sweep efficiency for most reservoirs.

Repeatability is essential for successful time-lapse or 4D processing. This will confirm the changes are due to the changes resulting from the waterflood rather than changes in the way the repeat survey was conducted.

Waterflood monitoring requires differentiation between segregated flow regimes (implying the existence of a flood front) and diffuse flow regimes. For segregated flow, the situation is simple, and amplitude data might be sufficient. For diffuse flow, however, proper saturation mapping is required, implying the need for acoustic impedance.

The feasibility and applicability of 4D for any given application are a function of a number of key variables:

- Acoustic properties of the reservoir fluids and rocks.
- Fluid compressibility contrast: A compressible fluid (oil, gas) replaced by incompressible brine will give a large 4D seismic effect that is easier to detect. Hence, light oils (or gas) and saline formation water are conducive to large 4D effects. Heavy oils and fresh water do not lend themselves so well to 4D.
- Rock compressibility: Reservoir compressibility is a function of rock and fluid compressibility. If rock compressibility is large, fluid fill changes will result in relatively large 4D effects. Very incompressible rocks render small 4D effects.

Good-quality seismic is necessary to justify a 4D seismic survey. Therefore, the technique is commonly used in offshore settings, where seismic quality is usually good. Sometimes, reservoirs lying below salt have poorer-quality responses that make them less attractive 4D candidates. The presence of large gas caps can also impair seismic quality. In general, 4D seismic surveys are less commonly used in onshore fields because the quality of seismic data is sometimes lower in such settings.

In offshore settings, seismic surveys have traditionally been obtained using streamers. A seismic-survey vessel tows streamers suspended below the surface that carry hydrophones. Sound waves are transmitted from the vessel using a compressed air gun, and the reflections from the different layers of rock are collected by the hydrophones located along the seismic streamers.

Many fields now use permanent arrays of geophones and hydrophones on the seabed that are connected using ocean-bottom cable (OBC) or by using nodes (Nakayama et al. 2012). These systems were originally introduced to enable surveys in areas where obstructions, such as the production platform, would impair accessibility for ships towing seismic streamers. They enable more-complete surveys of the field. They also facilitate more-frequent 4D field surveys and enable the acquisition of wide-azimuth and high-fold data sets.

The use of OBC typically entails an increased upfront cost, and as a result, streamer-based surveys probably still constitute the majority of applications. However, the improved functionality and flexibility afforded by OBC acquisition have clearly resulted in their increased deployment. Furthermore, OBC might be less costly than nonpermanent systems if regular monitoring is a requirement (a likely scenario for waterflood optimization).

With either nodes or cables, the key 4D advantage of seabed acquisition over towed-streamer acquisition is the greater control over receiver positioning. This gives improved repeatability between different survey vintages. Life-of-field arrays also enable improved consistency between different surveys. Seabed surveys can also provide multicomponent data, and, with an appropriate source grid, dense, wide-azimuth sampling. Such coverage can provide a wealth of additional information, enabling more extensive 4D analysis. These additional attributes potentially identify changes resulting from waterflood that might not be identifiable from conventional surface streamer surveys.

There are a number of variants of seabed-type systems (Effiom 2016). A complete trenched array can be installed on the seabed, and when data are required, a source boat is deployed to the field. Information can be obtained quickly with this system, but there is an appreciable upfront cost. The ocean-bottom node system entails the deployment of nodes through the use of remotely operated vehicles. This system can be deployed in fields with appreciable amounts of subsea infrastructure that might impair the deployment of a conventional ocean-bottom system. A source vessel is fitted with an air gun to provide the required compressed-air shots into the water.

The primary concern with this system is that, although high-quality data can be obtained, it is a time-consuming deployment methodology involving a high cost for each deployment. Indeed, the cost of a single survey might be half the cost of installing a full seabed array. This makes this option difficult to justify in fields where a large number of surveys might be planned.

There are two additional options to consider in relation to 4D acquisition. The first is termed i4D (Stammeijer et al. 2013). It is an option that provides a smaller-scale, high-quality-data survey focused on an individual injection well or a specific area of the field. It can be conducted between full-field surveys to better understand a key area of the flood displacement. In an example from the Gulf of Mexico, an i4D survey obtained using fewer than 25% of the nodes used in a full-field survey provided quick information regarding the location of out-of-zone injection (Effiom 2016) (Fig. 17).

Distributed acoustic sensing vertical seismic profiling enables high-resolution data acquisition by means of a fiber-optic cable installed in the tubing of wells. This significantly reduces the footprint associated with 4D acquisition. The seismic energy is again provided by a source vessel, which is the primary cost associated with this option. The primary drawback is that it is limited to the acquisition of data in a 2-km area around the well in question.

A relatively early example of the value of 4D surveys in a waterflood setting was Norway's Draugen Field (Mikkelsen et al. 2005). Fig. 18 shows the seismic amplitude maps from 4D surveys conducted at Draugen in 1998, 2001, and 2004. Increased shading indicates increasing water saturations. In this field, two clusters of injectors drive the flood from the northern and southern flanks of the field to crestal producers in the middle. In addition to the water movement from the northern and

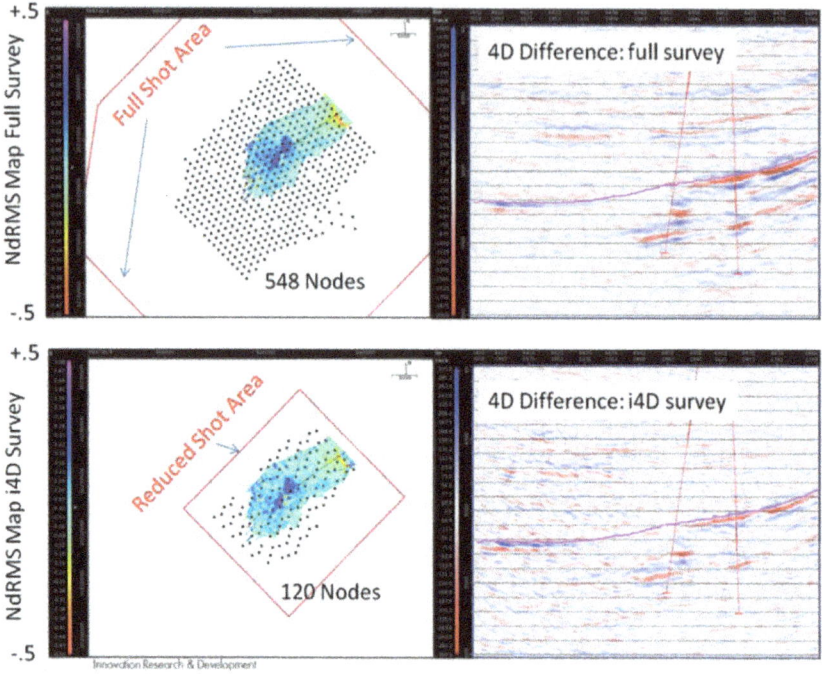

Fig. 17—Identification of out-of-zone injection using i4D (Effiom 2016).

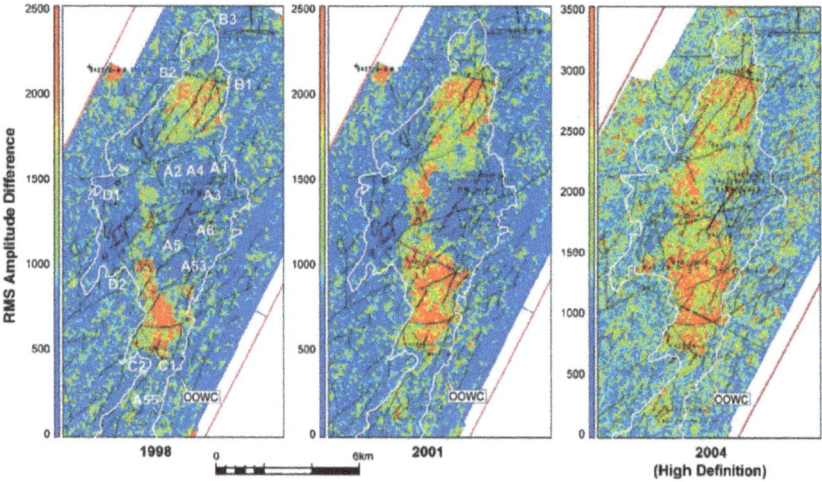

Fig. 18—Seismic amplitude maps from Draugen 4D surveys (Mikkelsen et al. 2005). RMS = root mean square; OOWC = original oil/water contact.

southern injector clusters, there is also water influx from the west, which had not originally been anticipated.

The movement of the aquifer water from the underlying Garn Formation into the producing Rogn Reservoir required a modification of the understanding of fault transmissibility. This improved the understanding of the reservoir and led to a more representative reservoir model. The saturation profiles in the 1998 survey demanded that a planned infill well to the west be relocated to avoid early water production. It was subsequently relocated farther to the north and east (and delivered record rates).

The 2001 data showed that the water influx was located much closer to the northern platform wells than had been expected. It also showed that there was an imbalance in the flood fronts between the south and the north. The accelerated water movement from the north identified the need to bring forward planned scaling treatments to safeguard production. Because of the proximity of the water fronts, the northern injectors were temporarily shut in to enable the treatments to be performed before breakthrough occurred.

Another good example of how 4D surveys can help understand sweep and its impacts is Australia's Enfield Field (Medd et al. 2010). A baseline survey was obtained in 2004, and then a 4D survey was conducted shortly after production began in 2007. This provided a clear 4D signal and early indications of field behavior. It provided the justification for updip injection and also highlighted some compartmentalization resulting from faulting in the eastern part of the field.

An additional survey in 2008 then provided the justification for a new producer/ injector pair in an area that had been overpressured but not swept. This survey was also needed to understand the lack of support experienced by a producer in the western part of the field (ENA02). A subtraction of the 2007 survey data from the 2008 data at the upper reservoir level showed a clear softening response around this producer and identified gas breakout resulting from a lack of pressure support (Fig. 19). The area to the north of this well seemed to show a clear barrier to flow that prevented support being felt from the ENB02 injector. The nature of that feature was uncertain because there was no evidence of faulting on the seismic, so it might have been a shale-filled channel or a subseismic fault. Either way, the data clearly demonstrated that water from ENB02 was being diverted to the east and into the heel of the ENA01 producer.

The 4D data suggested that the solution to the problem of providing support to ENA02 was to sidetrack the ENC01 injector to the north of the adjacent east–west fault. This action resulted in a doubling of production from ENA02.

4D seismic showed the baffle to be absent at the lower reservoir level. However, although water can flow in the lower level, the data suggest that, because no support is felt in the upper level, there must be some local stratigraphic barrier between these intervals in this part of the field.

The primary objective of 4D surveys is to understand the reservoir sweep in the target waterflood reservoir. However, it is also possible to conduct an analysis on the seismic reflections from the overburden because these will also be present in the obtained data set. This means that any injection that results in a loss of containment, such as induced fracturing resulting in out-of-zone injection, will also result in a 4D signal being observed in any layers above the target reservoir that receive any injection water.

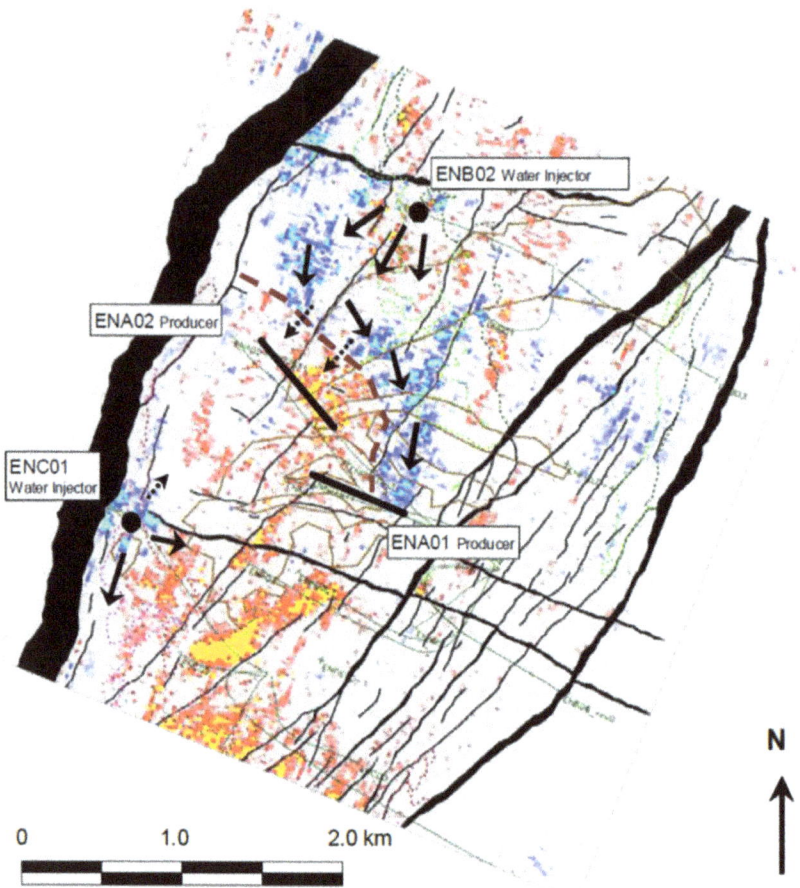

Fig. 19—4D response at Enfield (Medd et al. 2010).

In the early days of 4D deployments, one of the primary challenges was how to provide an economic justification for the appreciable cost involved. The fact that the value of 4D in so many recent field deployments has been completely self-evident means this has almost become standard technology for waterflood deployments in a number of major oil companies, provided that the seismic quality is adequate. Nevertheless, there are some companies in which the uptake has been much slower.

One of the critical issues to address in assessing whether 4D surveys might be an appropriate surveillance technique to manage waterflood is whether the quality of the seismic data supports adequate granularity in the 4D response. There then needs to be an assessment of the different ways in which value might be derived. These could include

- Infill-well location optimization: 4D surveys can identify the location of swept areas. This enables not only the targeting of infill wells into unswept locations but also enables their placement so that they are as far as possible from swept areas. This enables low water-cut production to be delivered for as long as possible.

- Remediation of sweep problems resulting from baffles/barriers: As the Enfield example shows, it is possible to understand the impact of reservoir barriers even in cases where those barriers might not be visible. It is also possible to optimize the appropriate remediation option. Identifying areas of compartmentalization that might not be swept by the existing flood can also be achieved. In some cases, this could provide a justification for additional injector/producer pairs.
- Reduction in well investments: Enabling optimized waterflood sweep might allow the development to be achieved by a smaller well stock than would otherwise be needed, thereby reducing the required capital expenditure.
- Economics: Overall field economics might be improved by the reduction in well stock or through other benefits such as accelerating production and recovery, extending field plateau rates, minimizing field declines, or extending field life to delay eventual abandonment.
- Optimization of scale-treatment timing: A detailed understanding of the location of flood fronts can be used to quantify the timing of injection-water breakthrough. In cases where scale squeezes are required in producing wells, this enables the optimized timing of such treatments, which can consequently secure and optimize production.
- Improvement to available reservoir models: The data from 4D surveys can be used to improve the available history match using a workflow for the quantitative incorporation of 4D seismic and production data into the reservoir simulation model (Landa and Kumar 2011). This can then improve the quality of the future production forecasts.

The value of 4D seismic has now become so widely accepted that some companies use it almost as a standard, particularly in deepwater environments. Even so, a value of information exercise could still be appropriate to provide the justification for a 4D survey, such as was performed in Brazil's Marlim Field (Steagall et al. 2005).

3.4.4 Electromagnetic Surveys. Electromagnetic (EM) surveys use magnetic dipoles as source and receiver. They couple inductively to the formation to provide an image of the resistivity distribution between wells. The interwell measurements are taken by placing a transmitter in one well while a second well detects those transmissions by means of an array of induction coil (**Fig. 20**). Because waterflood induces saturation changes within the reservoir, there will naturally be changes in resistivity as the flood front progresses. In principle, those changes can be detected by means of such EM surveys.

The measurements are fitted to calculated data from a numerical model using an inversion procedure. Although the inversion process results in nonunique models, it is normally possible to exercise reasonable model constraints using existing field data as a guide to generate a best-fit model.

This methodology is not used as frequently as 4D surveys. The presence of steel casing results in very large attenuation effects and phase shifts in the EM data, which increase with increasing frequency. Such effects could, in principle, be estimated from the EM response. However, in practice, the presence of steel casing significantly reduces the sensitivity to the formation, especially near the well.

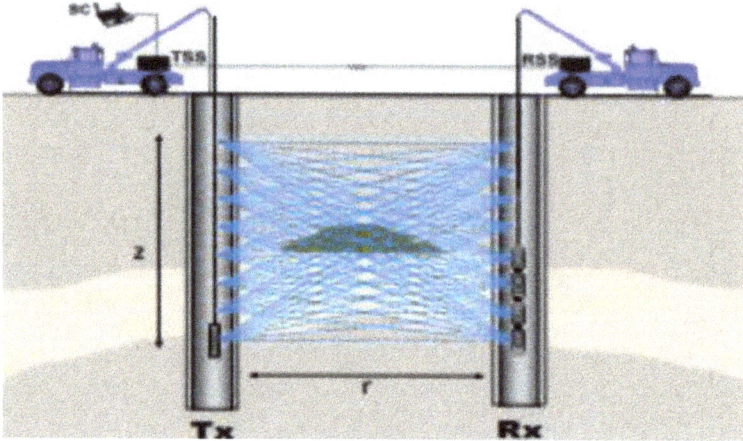

Fig. 20—Crosswell EM survey (Mishra et al. 2019).

Shallow transient EM surveys that measure voltage decay (secondary EM field) in the subsurface in response to the pulse-like turnoff of a transmitter current (primary EM field) have been used to investigate waterflood performance in a carbonate reservoir in the Irkutsk region of Russia (Sibilev et al. 2019). However, the granularity of subsequent data does not appear to be at the same level as that normally achievable through 4D surveys.

Mishra et al. (2019) describe a workflow to integrate crosswell EM data to streamline simulation to improve the predictability of the available dynamic model.

3.4.5 Tracers. Tracers are typically chemicals that can be injected into injection wells, although there are other, more esoteric deployment means such as by means of impregnation into gravel-pack completions. Such chemicals need to be detectable at low concentrations because the aim is to detect them in production wells. This enables the quantification of reservoir transit times so that information regarding reservoir flow paths can be derived. If a different tracer is used in each injection well, the derived information regarding reservoir sweep patterns can be quite detailed.

For a tracer to be used in a waterflood, it must be able to transit the reservoir without reacting with materials in the reservoir, and it must also be chemically stable at the temperature and pressure conditions in the reservoir (Tayyib et al. 2019). Historically, tracer studies have tended to be conducted with radioactive chemicals. A tritium tracer study was conducted at Prudhoe Bay, where the breakthrough patterns of the tracer suggested the vertical permeability within the reservoir was somewhat lower than previously assumed (which was good for recovery) (Nitzberg and Broman 1992). It also showed that faults perpendicular to water movement were not sealing and could therefore act as flood short circuits.

The benefit of radioactive tracers is that the radioactive isotope can be incorporated into a molecule chemically equivalent to one already involved in the process. Tritium (an isotope of hydrogen) can be incorporated into a water molecule. Thus, it can be reasonably assumed the isotope would transit the reservoir in exactly the same manner as other water molecules (the only difference is that tritiated water is slightly heavier). A second benefit is that radioactive molecules can be detected at lower concentrations than alternative tracers.

Of course, the primary drawback of radioactive tracers is the health, safety, and environment (HSE) concerns associated with their use. Because the detection limits for alternative tracer options are now similar to those achievable with radioactive tracers, the radioactive tracers have been almost universally replaced. Most studies now use fluorinated benzoic acids (FBAs) because very small changes to the chemistry are easily made such that it is possible to inject a uniquely identifiable product into each injection well and the molecules can be detected at very low limits using gas chromatograph mass spectroscopy. (**Fig. 21** highlights a few of the wide range of molecules that can be used.) This chemistry also benefits from being stable at reservoir conditions, being quite inert to reaction and adsorption during the reservoir transit, and not exhibiting partitioning into other phases.

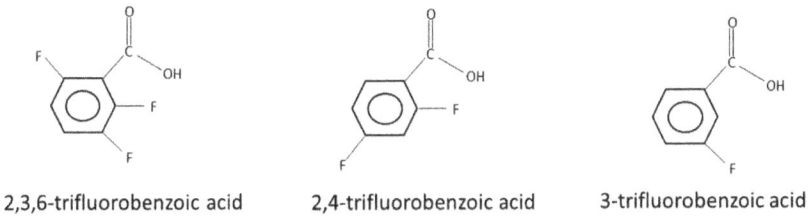

2,3,6-trifluorobenzoic acid   2,4-trifluorobenzoic acid   3-trifluorobenzoic acid

Fig. 21—FBA-tracer chemistry.

Tracer programs require regular sampling of water from producing wells. To minimize the surveillance requirements, samples are regularly taken but only analyzed periodically. When a sample containing tracer is identified, a more focused sample-analysis program is used to pin down the transit times more accurately.

The value of tracer programs can include

- Quantification of fault-block communication
- Identification of reservoir anisotropy
- Location of thief zones resulting from channeling/fractures
- Improvement in the understanding of sweep pathways, leading to potential infill and/or improved injection optimization
- Improvement in the understanding of reservoir flow pathways, which can be expected to improve the accuracy of H_2S forecasts
- Calibration of simulation models using tracer breakthrough time
- Calculation of volumetric sweep efficiency using average tracer transit time between each injector-to-producer pair
- Identification of multiple porosity/permeability systems and their respective fractional volumes

In their simplest form, tracer data can be used to elucidate reservoir breakthrough times (usually the tracer is best injected at the outset of the project). However, most simulators feature the ability to incorporate tracer transport data, so these can be incorporated in the same manner as other production data. An alternative to the laborious incorporation of data into simulation is to perform what is known as residence time distribution analysis. This is the distribution of the times it takes a population of tracer particles to travel through a medium. Where an injector is connected to multiple producers, the residence time distributions between each injector and producer can be defined. This allows important information regarding the geometry and flow in a system to be obtained. One example of how residence time distribution analysis can give valuable information is provided in a study conducted in the Matzen Field (Huseby et al. 2016). **Fig. 22** shows an outcome from this study, with the yellow arrows representing the flow of water toward the producers, the widths of each arrow being proportional to the amount of tracer produced in each of the producers, and the areas of the ellipses corresponding to the swept areas for each injector/producer pair.

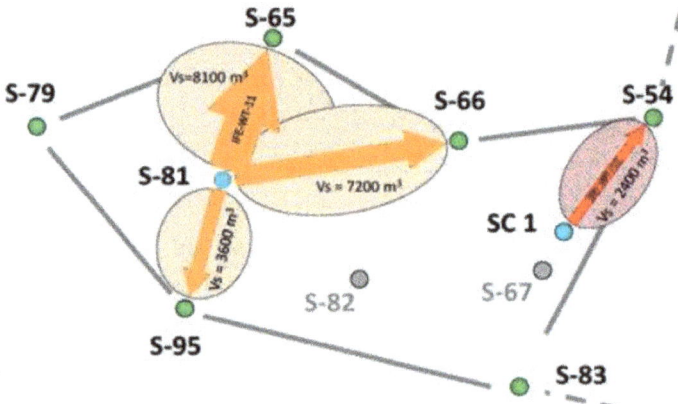

Fig. 22—Results from residence time distribution analysis of tracer data in the Matzen Field (Tayyib et al. 2019, after Huseby et al. 2016).

The improved detection capability with tracers has now reached the stage that it has even been suggested that real-time surveillance of tracer breakthroughs could be appropriate (Nikjoo 2019).

An additional type of tracer test is one in which partitioning tracers are used (Viig et al. 2013). In such tests, the performance of the partitioning tracers is compared to conventional FBA tracers to derive an estimation of remaining fluid saturations.

This is effectively an alternative to a relatively established technique known as the single-well chemical tracer test (SWCTT). This technique uses periods of injection, shut-in, and production to evaluate residual oil saturation. Most reported SWCTT operations use ethyl acetate as the primary reacting tracer in addition to n-propyl alcohol and isopropyl alcohol as cover tracer and mass balance tracer, respectively. The interpretation of these tests, which record the concentration of returning tracer, is based on chromatography theory, with the partitioning species being delayed by the presence of oil. Consequently, the delay in production of partitioning ethyl acetate is compared to that of the alcohols to establish the oil saturation. One such test was performed in a West African onshore field to assess the viability of surfactant or low-salinity flooding as an alternative to existing seawater flooding (Callegaro et al. 2014).

A more sophisticated tracer program is possible, in which the tracer is incorporated in the injector or producer completion (the tracers can be integrated into sand screens, liners, tubing) when the injection well is initially completed. The tracers are then removed by the flow, thus enabling a rigorous understanding of reservoir flow pathways. One example of such a deployment is the North Amethyst Field, located off the east coast of Newfoundland and Labrador, Canada (Montes et al. 2013). The horizontal wells used inflow control devices (ICDs) to equalize the flow. The subsea development prevented the application of traditional means to establish adequate sweep. Therefore, tracer rods and filaments were designed to deploy oil and water tracers. The shape of these systems was also designed to fit into the voids of the wire-wrapped screens integrated with an ICD (**Fig. 23**). In this case, capture of all the installed tracers demonstrated that the installed ICD design was able to facilitate a full horizontal contribution and was therefore promoting good waterflood sweep.

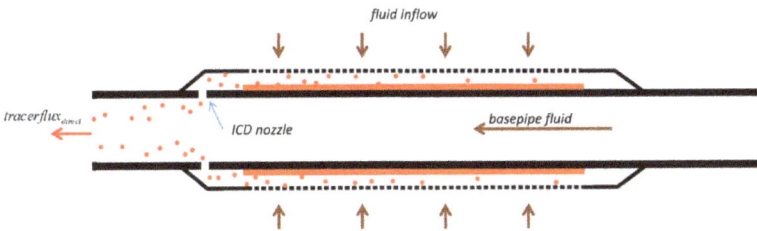

Fig. 23—Screen joint with fluid inflow and tracer system wetting through the screen section (Montes et al. 2013).

Another possible option in some cases is to use natural tracers that are part of ongoing flood operations. One example for such a tracer is to analyze producing wells for biocide when tetrakis(hydroxymethyl)phosphonium sulfate (THPS) is the biocide used for bacterial control in the injection system.

3.4.6 Interference Tests. Interference tests are used to determine whether two or more wells are in pressure communication. When communication exists, the tests are used to provide estimates of permeability and porosity in the vicinity of the tested wells. In the interference test, fluid is produced from, or injected into, at least one well (the active well) while observing the pressure response in at least one other well (the observation well).

Assuming the active well to be a producer, as it starts production it experiences a pressure decline, and the pressure in the observation well then begins to respond after a time lag. The magnitude and timing of the deviation in pressure response at the observation well depend on reservoir-rock and fluid properties in the vicinity of the active and observation wells.

An interference test was conducted between an injector and a producer in the turbiditic Girassol Field before first oil from the field (Retail et al. 2002). This was performed because of uncertainties associated with connectivity in the field resulting from concerns about the risk of sealing faults. It raised the possibility that the number of wells needed to maximize recovery might be greater than originally planned. In this case, a pulse was applied to the injector, with the offset producer being used as an observation well because the test was conducted before the arrival of the floating production, storage, and offloading vessel. This therefore avoided the need for the flaring of any produced fluids. The targeted pressure increase of 0.15 bar at the production well was observed within 40 hours, suggesting excellent reservoir communication and that two faults located between the wells were not sealing. This valuable information on fault transmissibility was used to modify the flood pattern and, as a result, the number of wells needed for the development was reduced by one, translating into savings of USD 18 million.

Development of deepwater fields presents a double-edged sword in terms of the need for interference testing. Although such tests can be expensive because of the rig time required to perform them before the development is finalized, they can provide invaluable data about the reservoir and the degree of connectivity within it. Such tests are therefore always likely to find a place in such settings, but testing should be carefully considered to ensure objectives are met at the lowest cost. Therefore, when exploration and appraisal wells are abandoned or suspended, there could be great value to leaving pressure gauges in those wells to enable an understanding of the connectivity to be derived from production or injection tests (Zubarev et al. 2019).

3.4.7 Streamline Modeling. Streamline modeling complements standard finite-difference simulation. The underlying process in streamline-based flow simulation decomposes a 3D displacement into a series of 1D displacements along streamlines. Fluid transport occurs along this streamline-based grid, rather than the underlying Cartesian grid. Streamline simulations offer two unique advantages compared to conventional finite-difference simulation:

1. Improved computational speed: Because streamlines are 1D systems, it is easy to efficiently solve the transport problem. As a result, the run times can be dramatically decreased.
2. Improved understanding of reservoir flow: The visualization of streamlines improves the user's understanding of the displacement processes in the reservoir (e.g., sweep). In addition, streamlines enable the calculation of additional data related to the volumes connected to each well (e.g., influence of injectors on producers, drainage area).

A good example of the use of streamline simulation in waterflood optimization is Alaska's Prudhoe Bay Field (Grinestaff and Caffrey 2000), where streamline simulation identified flood short circuiting (**Fig. 24**). Injected fluids were fluxing to producers several patterns away and large amounts of injected water were leaving the designated waterflood area. This problem was resolved through drilling new wells as well as using water shutoff treatments. As a result, the underlying oil decline reduced from 22 to 11%.

Another good example is Oman's Lekhwair Field (Giordano et al. 2007), where streamline simulation showed that much of the injected water was traveling through the underlying aquifer without sweeping any oil (**Fig. 25**). This short circuit was subsequently resolved through the use of plugs installed in the impacted injectors.

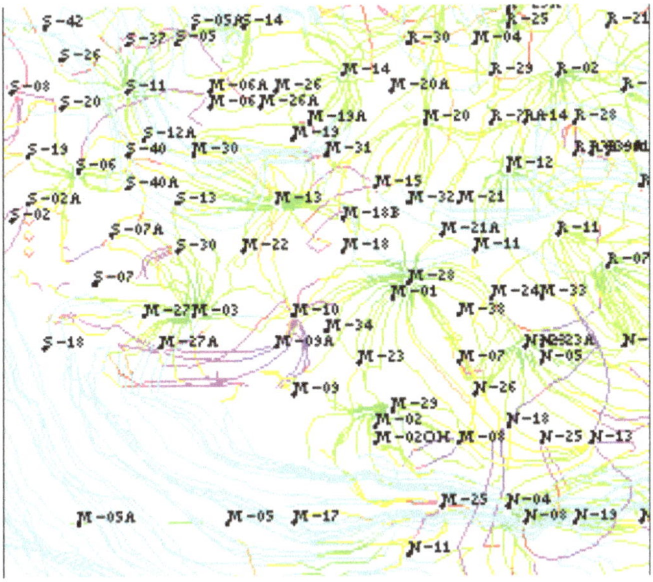

Fig. 24—Reservoir streamlines at Prudhoe Bay (Grinestaff and Caffrey 2000).

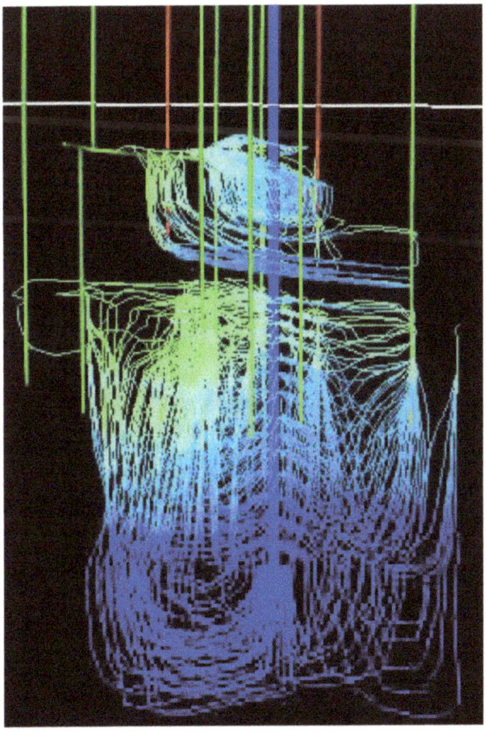

Fig. 25—Streamlines around an injector, Lekhwair Field (Giordano et al. 2007).

Another example, also from Oman but this time in a sandstone reservoir, is the Thuleilat Field (Naguib et al. 2006). In this case, the short circuits were lateral in a pattern flood, and it was possible to divert sweep patterns by shutting in or modifying production from producers where flood short circuits occurred.

3.4.8 Capacitance Resistance Modeling. Capacitance resistance modeling (CRM) is an analytical technique that can facilitate a rapid evaluation of waterflood performance (Sayarpour et al. 2008). In the last few years, it has begun to generate a reasonable amount of interest. Rate variation at an injector introduces a signal that can be expected to induce a corresponding response at one or more producing wells. Consequently, CRM uses production and injection rate data (as well as downhole pressure data, if available) to calibrate the model against a specific reservoir. After this, the model can be used for predictive purposes. In other words, it is a simplified material-balance model.

It only uses production and injection data to predict performance, which allows for simplicity and speed of calculation. After a model has been calibrated against the production/injection data, it can be used to optimize future production by allocating the available water-injection volumes. Although CRM-derived interwell connectivities can sound like they strongly resemble streamline allocation factors, there can be noticeable differences. This is because CRM-derived interwell connectivities are principally related to the pressure support, while streamline allocation factors are related to the fraction of injected fluid flowing toward a producer. Such subtle differences can be very useful in understanding and therefore optimizing waterflood performance.

The application of CRM modeling to a south Oman waterflooded reservoir suggested, perhaps somewhat optimistically, that the use of this technique would lead to an improvement in oil production of nearly 30% (Al Saidi et al. 2015). It has also been used to better understand connectivity between injectors and producers in a field in the Niger Delta. The results of Nwogu et al. (2019) identified strong interwell connectivity between active producers and two idle injectors, and this finding supported a revised geologic interpretation of the reservoir. The injectors in question had previously been shut in because of a belief that shale was negatively impacting connectivity and performance. Consequently, the injectors were reopened, and a good pressure and production response was observed. This subsequently allowed work on a planned new injector to be shelved, resulting in USD 20 million savings.

4. Data Analysis

4.1 Well and Pattern Reviews. Waterflood management is often facilitated by the use of regular, integrated reviews of well and field data that can be used to identify opportunities for improvement. It might be possible to review small fields as a whole on a regular basis. In contrast, it is likely that larger fields will be segmented, with the different segments being reviewed regularly in sequence. To ensure all the available relevant information is used in the discussions, it is important that all disciplines be represented in such events.

In fields with a large number of wells to be analyzed, the use of a so-called after-before-compare (ABC) plot (**Fig. 26**) could be useful to diagnose wells that might need attention (Terrado et al. 2007). This plot compares oil and water production data between specific dates. The *x*-axis compares the current water rate with the previous water rate, and the *y*-axis makes the same comparison for the oil rates. Most wells lie close to the 1:1 area, and they are unlikely to need any attention because there has been little change between the two dates. Wells that need further investigation include those still on the 45° slope line but that are falling significantly below the 1:1 coordinate point, indicating an appreciable rate decrease, and those to the lower right of the 45° slope, indicating an increase in producing water cut.

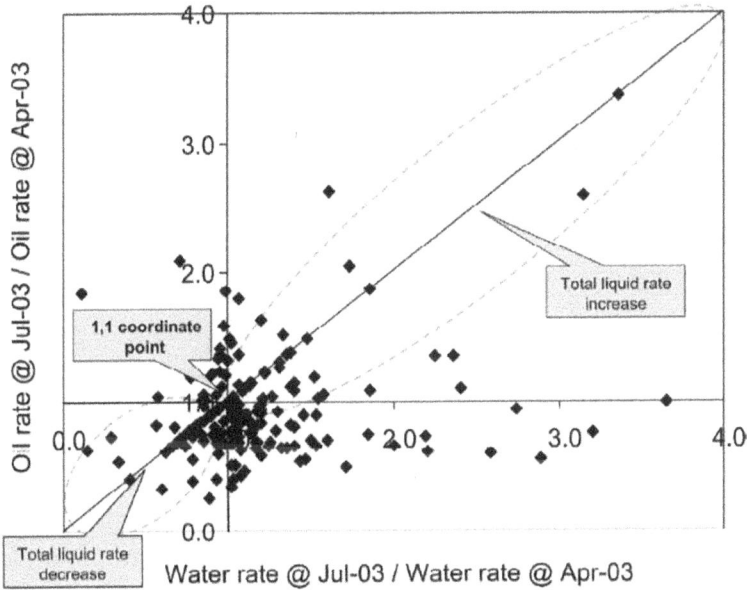

Fig. 26—ABC plot (Terrado et al. 2007).

Such reviews can be facilitated by software that is able to collate and present the data in formats that make interpretation of trends easier. Many such tools are available within the industry. One such example has been applied in the Western Siberian Salym fields (Mijnarends et al. 2015), where nearly 1,000 wells are used to develop the stacked sand sequences in a regular pattern arrangement. For waterflood management purposes, wells are arranged in groups of 20–50 based on the impacts of the geology and development history. Large amounts of data have been collected to ensure that the flood performance is understood. Routine data measurements include

- Continuous monitoring of ESP lift performance.
- Automatic testing of gross production five times per month and basic sediment and water measurement performed weekly.

- Continuous metering of injection-well volumes using orifice plates.
- Monitoring of pressures by a routine program of flowing pressure surveys and pressure falloff tests, supplemented by openhole formation pressure tests. These data are used to build pressure maps that are updated monthly.
- Periodic logging exercises to update zonal allocations in both producers and injectors, which are initially based on permeability-thickness product.

To ensure maximum value is derived from these data, the operator invested heavily in its enterprise information architecture. It also built a back-end data infrastructure system that checks quality and storage capability before feeding into a host of different applications.

All wells are assigned to an injector-centered pattern and used in setting injection targets based on voidage replacement requirements and observed pressures. However, the individual patterns were found to be of limited value in overall flood management because many activities affect multiple patterns, and fluids often move over distances larger than individual patterns, which complicates the allocation process. Consequently, it was found that the impact of activities is better assessed at a higher level and the flood is best managed at the block rather than the pattern level. Issues flagged at that level are then investigated in more detail at the pattern level.

To efficiently manage the flood in such a large field, a dedicated software platform for flood surveillance and analysis was built to provide the following functionalities:

- Create an integrated database containing key geological and petrophysical production and well status.
- Handle data at well, pattern, block, and field levels.
- Provide a hard link to the reservoir simulation model that provides updated maps and streamlines for input into performance reviews.
- Determine streamline-based allocation factors and pattern and block recovery factors and sweep efficiencies.
- Provide a user interface giving access to
  - Layered maps displaying production and injection bubble maps.
  - Historical rate and pressure data at well, pattern, block, and field levels.
  - Diagnostic plots for use in pattern reviews.
  - Linkage of well performance to that of its eight nearest neighbors (see **Fig. 27**, for example).

The use of streamline-based water allocation factors in patterns has provided a much more realistic understanding of flood performance. **Fig. 28** compares pattern performance in a block using geometric water allocation factors, streamline-based allocation factors, and calculation from the simulation model. The geometric-based assessment suggests an unequal flood performance, but when fluid flux between patterns is taken into account, a much more balanced flood performance, believed to reflect the true position of the development, emerges.

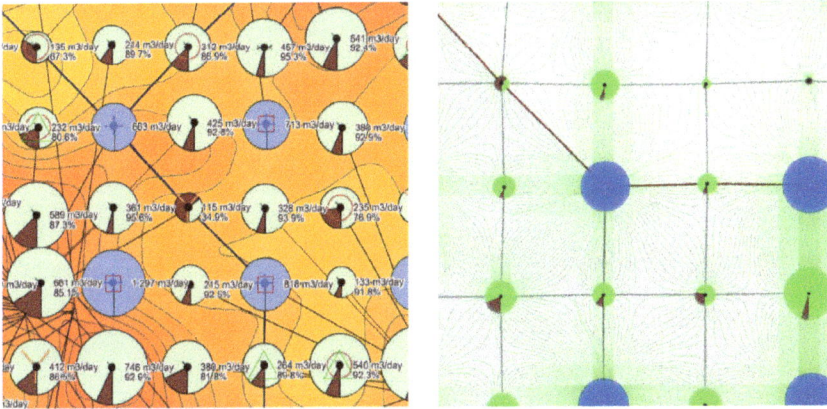

Fig. 27—Layered map and streamline map generated by waterflood management tool (Mijnarends et al. 2015).

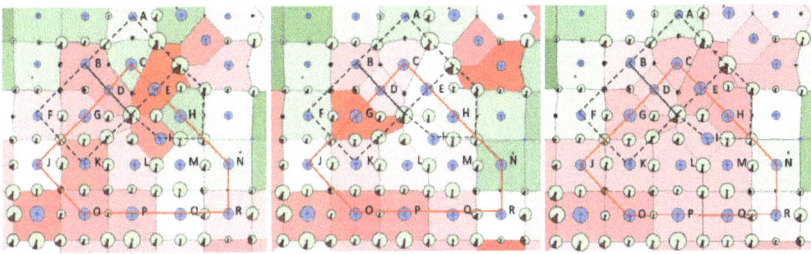

Fig. 28—Pattern maps of recovery based on different methods of water allocation (Mijnarends et al. 2015).

Overall, the tool provides a sound foundation for flood management in this large and complex field. It also provides the basis for well activities. This is a key issue for management of all pattern-based floods. In pattern floods in north Oman, this process was responsible for 5.4% of incremental oil production while safeguarding 23% of existing production and was therefore able to arrest the ongoing natural oil rate declines observed in these fields (Al-Bimani et al. 2006).

4.2 Diagnostic Plots. Routine data can often be manipulated in specific ways that provide useful pointers regarding flood performance. A number of such plots are identified in this section.

4.2.1 Chan Plot. An analysis of log-log plots of WOR against time, or GOR against time, showed that different characteristic trends were often seen for different water- and gas-breakthrough mechanisms (Chan 1995). The time derivatives of WOR and GOR were found to be capable of differentiating whether the well was experiencing water and gas coning, high-permeability-layer breakthrough, or near-wellbore channeling.

Fig. 29 shows that in a log-log plot of the WOR (rather than water cut) against time, there is a clear distinction between the trends for water breakthrough resulting from water coning and breakthrough by means of a high-permeability streak (channeling). For coning, the rate of the WOR increase is relatively slow and gradually approaches a constant value. For channeling, the water production from the breakthrough layer increases very quickly, so the WOR increases relatively fast.

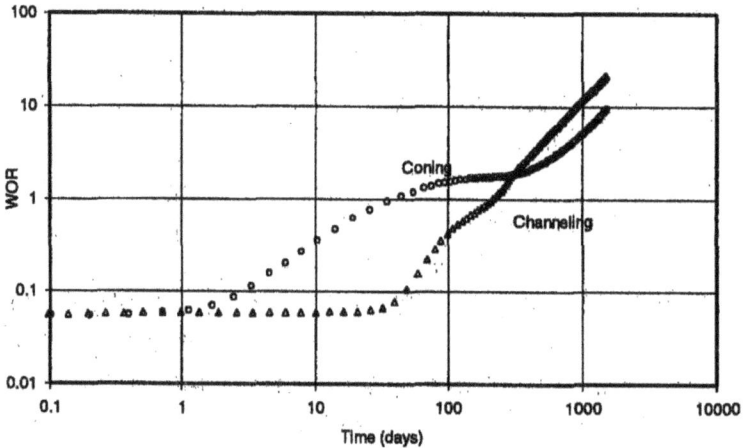

Fig. 29—WOR comparison of water channeling and coning (Chan 1995).

Sometimes, the time derivatives of WOR can be useful to differentiate coning from channeling. **Figs. 30 and 31** show the WOR and WOR′ derivatives for

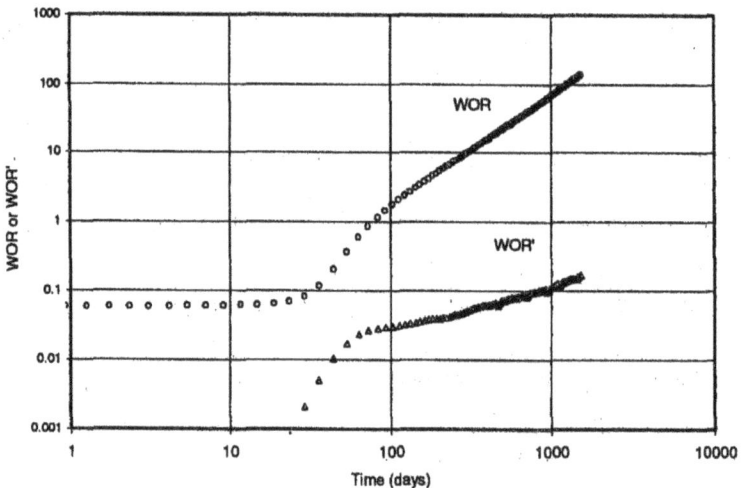

Fig. 30—Multilayer channeling WOR and derivative (Chan 1995).

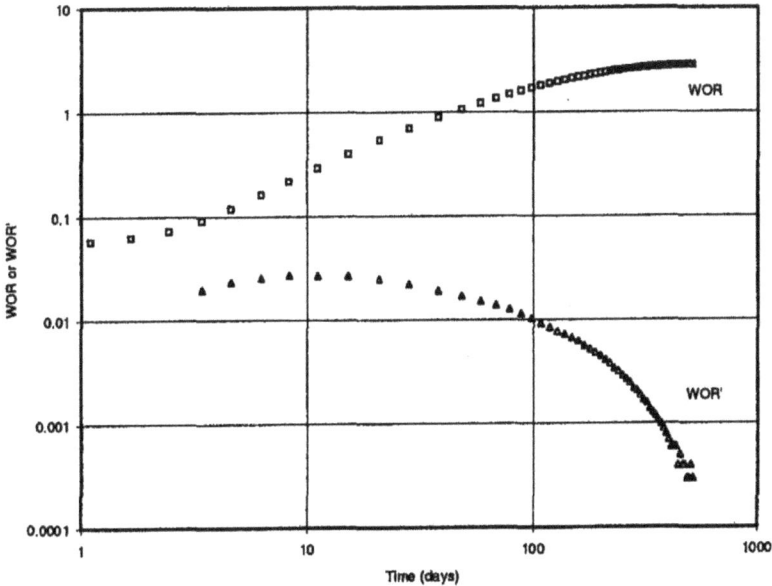

Fig. 31—Bottomwater coning WOR and derivative (Chan 1995).

channeling and coning, respectively. The WOR' (simple time derivative of WOR) shows a nearly constant positive slope for channeling and a changing negative slope for coning.

Although the original Chan plot is based on a time measurement, the data can be more diagnostic if the plot is based on cumulative fluid.

4.2.2 Recovery Efficiency. A plot of the achieved recovery factor against the amount of injected water, measured as a function of the oil volumes in place, can provide useful information regarding waterflood efficiency. The efficiency is expected to deteriorate after water breakthrough because of relative permeability effects. However, downward trends in recovery efficiency might also indicate that flood short-circuit pathways have opened up.

In **Fig. 32,** the y-axis = cumulative oil produced/stock-tank oil initially in place (STOIIP) and the x-axis = $(W_i \times B_w)/(\text{STOIIP} \times B_o)$, where B_w is the water formation volume factor and B_o is the oil formation volume factor.

Fig. 33 shows such a plot for 20 peripheral waterflood projects in the Denver Basin in the US. Although the recovery efficiencies in these projects appear to be unrealistically high, the shape of a number of these plots is the same as in Fig. 32, exhibiting reduced displacement efficiency as the flood progressively matures. However, a number of the fields exhibit an S-shaped recovery efficiency curve. This implies that efficiency is poor at first, before improving and then once again deteriorating. This effect is due to the presence of secondary gas in these fields at the start of the flood, resulting in an initial fill-up period as this gas is put back into

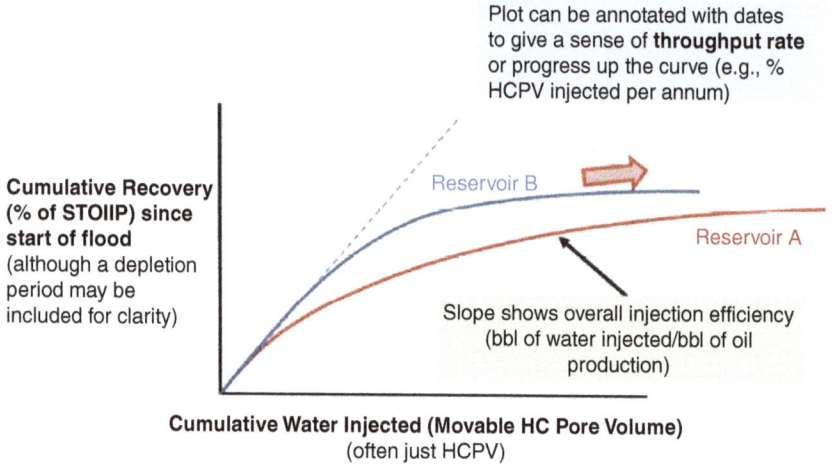

Fig. 32—Recovery efficiency (after German 2015). HC = hydrocarbon; HCPV = hydrocarbon pore volume.

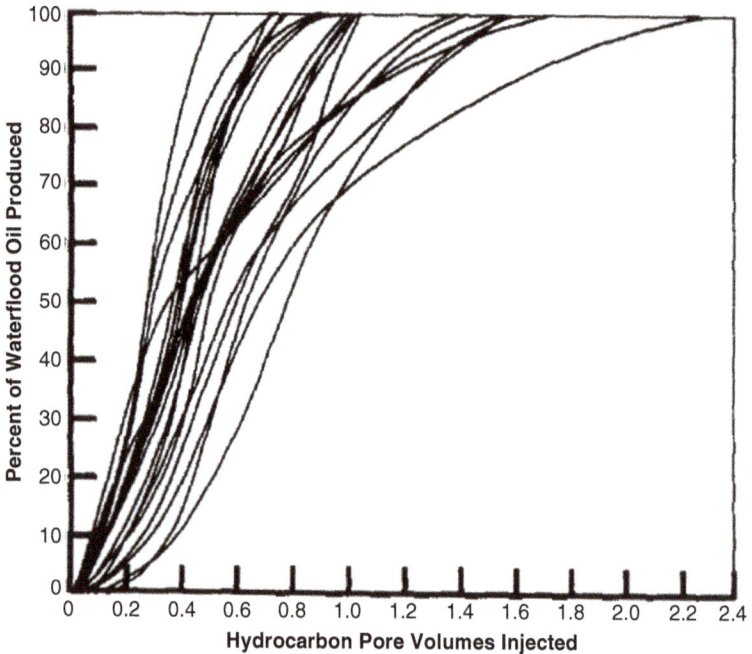

Fig. 33—Recovery efficiency in 20 Denver Basin waterfloods (Wayhan 1972).

solution. The recovery efficiency then improves as this period concludes, and from that point, these fields have recovery efficiency shapes that are the same as those of the other fields.

4.2.3 Voidage Replacement Ratio. A monthly plot of the voidage replacement ratio (VRR) helps to ascertain if enough water is being injected. (If there is some aquifer support, the VRR might not need to be as high as 1, but there will still be a need for a defined VRR requirement.) As shown in **Fig. 34,** it could be useful to plot VRR against the production rate. In this example, from the El Trapial Field in Argentina, a clear linkage between the VRR and oil production can be observed.

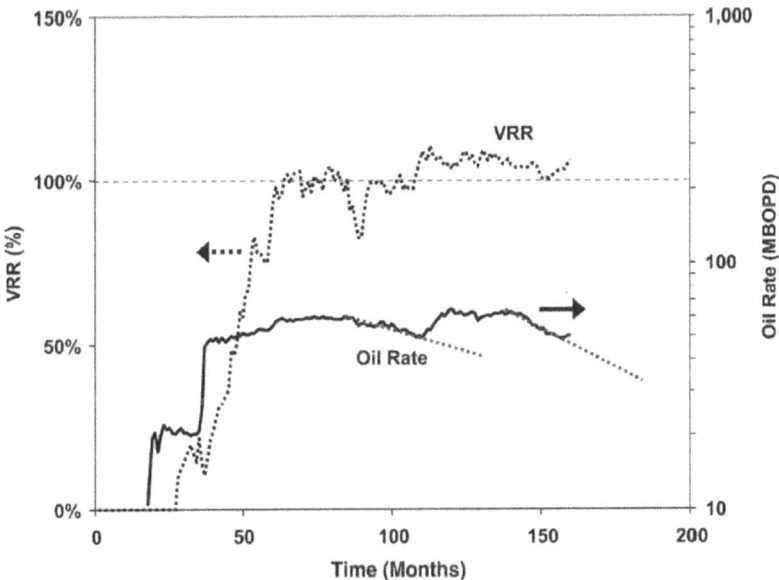

Fig. 34—VRR and production rate (Terrado et al. 2007).

It is useful to assess the effect the VRR has on reservoir pressure. If the required voidage is being exceeded and there is no increase in the reservoir pressure, this might indicate that some of the injected water is being lost or that there is some out-of-zone injection.

In addition to plotting the monthly VRR, the cumulative VRR should also be plotted to ascertain if sufficient water has been injected over the project life or whether overall water injection might need to be caught up.

Historically, many waterfloods had to commence with VRRs significantly greater than 1 to repressurize the field after an initial period of depletion. However, in some fields, there can be a need to inject with a VRR greater than 1 even in the longer term. This could occur where there is peripheral injection in a regionally connected aquifer, for example. Of course, any out-of-zone injection would also typically require high VRRs.

4.2.4 Productivity and Injectivity Index. Plots of productivity and injectivity index (B/D/psi) can be useful to demonstrate trends in performance of both producer and injector wells. Plots of this parameter against skin or permeability × height can be used to identify stimulation candidates. When the plot against permeability × height suggests a problem for an injector but no problem is indicated by the skin, it is possible that the well is injecting into a small compartment.

4.3 Yang Plot. The Yang plot is a diagnostic methodology based on the analytical solution of oil fractional flow against displacement time for mature waterflood reservoirs (Yang 2008). This analysis defines the oil fraction function, Y, as

$$Y = \left(\frac{E_v}{B}\right)\frac{1}{t_D}, \quad \dots\dots\dots\dots\dots\dots\dots\dots\dots\dots\dots\dots\dots\dots\dots\dots (3)$$

where B is the relative permeability ratio parameter, E_v is the volumetric sweep efficiency, and t_D is the ratio of cumulative liquid production (Q_L) to the total pore volume of the pattern area (including both swept and unswept areas).

In Eq. 3, B and E_v are unknown parameters. The relative permeability curves can be determined from laboratory measurements or a history match in a reservoir simulation exercise. When B is known, the volumetric sweep can be quantified. **Fig. 35** shows historical production data plotted on a reciprocal time scale

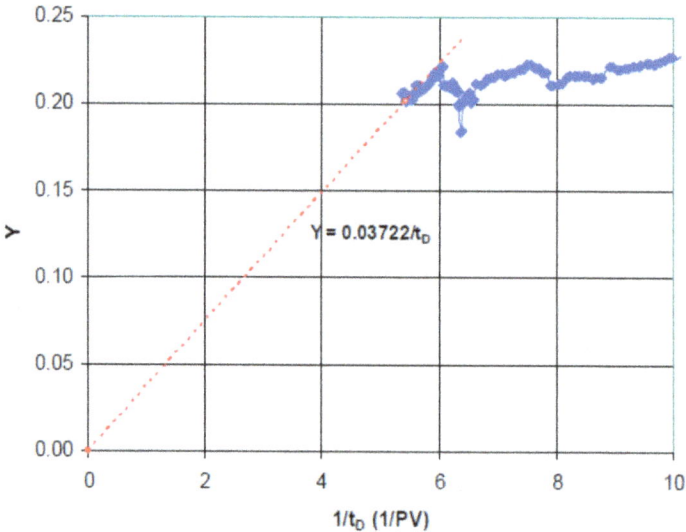

Fig. 35—Calculation of volumetric sweep from Yang methodology (Yang 2008). PV = pore volume.

for a field. When $1/t_D = 6$, drilling was complete, so the development could be considered mature. The slope of this part of the plot gives $E_v/B = 0.03722$, and from the history match, $B = 16.91797$. This consequently gives the volumetric sweep as 0.553.

An example from the Huntington Beach Field in California illustrates how the Yang model is used to define waterflood maturity, evaluate performance, and calculate ultimate recovery. **Fig. 36** shows Y, with an emphasis on the mature-flood period. Infill will influence the volumetric sweep, but when production stabilizes after the infill is complete, the plot of Y against t_D on the log-log scale resumes a straight line with a slope of -1 and intercept E_v/B. Any operational change that influences the volumetric sweep will influence the intercept on the log-log plot, although the slope remains constant. Fig. 36 shows several changes that influence recovery, and those impacts can be observed using the plot. The incremental recovery impacts of each change can also be observed.

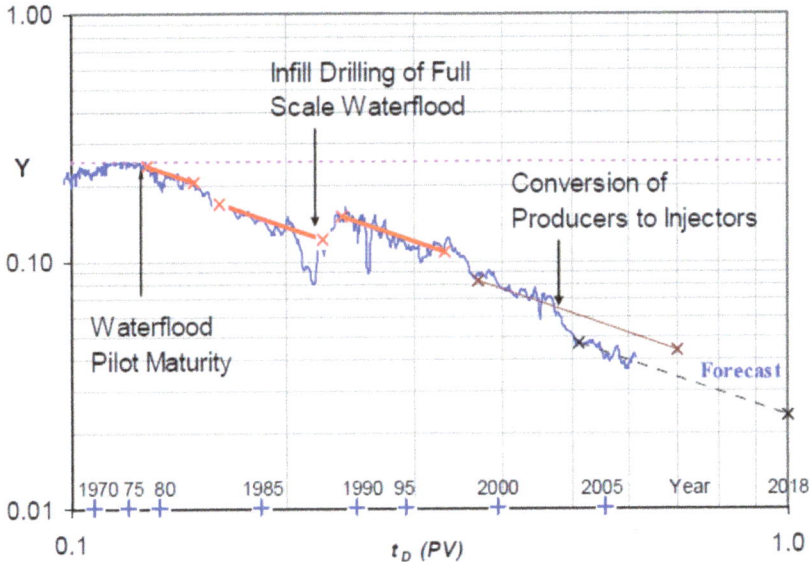

Fig. 36—Log-log Yang plot of Huntington Beach Field, California, USA (Yang 2008). PV = pore volume.

5. Remediation, Further Development, and Recovery Optimization

The number of degrees of freedom available for waterflood remediation are somewhat limited. One option is to go faster. However, the simplest option to achieve this—increase the injection pressure and simultaneously increase off-take from existing producers—is unlikely to improve performance unless the flood was not previously optimized in this respect and might be more likely to induce a flood short circuit. Well stimulation might be an option to increase throughput, as might the addition of injection or production capacity. Waterflood infill is another option to make the flood go faster. An alternative option is to go slower. This might be a viable strategy where fractured injection has induced flood short circuits.

Apart from this, the main options are to modify existing pressure sinks through the reservoir. Injected water has a tendency to follow established water pathways through the reservoir, and when a pressure gradient has become established in the reservoir, this tendency can become further exaggerated. Changing the pressure profiles will therefore assist in changing the established streamlines so that new oil, rather than existing water, starts to be swept. New wells, changing the operation mode of existing wells, or modifying the flow within both injection and production wells can all contribute to such changes. The means to effect such improvements will be discussed in more detail in the sections that follow.

It is also appropriate to constantly monitor the injection efficiency achieved by each injection well—that is, the volume of oil delivered from each barrel of injected water. During periods of injection constraint (such as can occur if an injection pump is unavailable), the swing list can be used to quantify which wells should remain onstream to result in production optimization.

5.1 Benchmarking. Appropriate benchmarking can be a useful tool to ascertain whether the existing development is doing everything possible to maximize recovery from the field. However, waterflood benchmarking is not a straightforward proposition because there are so many different factors that will influence recovery efficiency.

The more complex the waterflood environment (increased heterogeneity, heavier oil, lower permeability, and other factors), the more difficult it will be to deliver a high recovery efficiency. One way to approach this problem is to assess the various factors that comprise complexity in a waterflood context. Those factors can then be combined into a complexity index that can then be plotted against the recovery factor for different fields. Ideally, this approach will assign a relative importance to each of the complexity factors.

One case uses this approach to define the recovery that might be considered as being in the top quartile for a given level of complexity (Prelicz et al. 2014). The complexity is based on five categories:

- Pore scale displacement, which is expected to be influenced by factors such as wettability, saturation, and relative permeability
- Drainage/fault compartmentalization, which can be influenced by reservoir thickness, fault density, and the sealing capacity of those faults
- Areal sweep efficiency, which will be influenced by the lateral reservoir continuity, degree of heterogeneity, and presence of high-permeability streaks or fractures
- Vertical sweep efficiency, which can be influenced by a gas cap (and its size), vertical baffling or layering, and the thickness of the oil column in relation to the total reservoir thickness
- Economic and environmental factors, such as the location of the field, commercial terms, and the value of the oil itself

These data are compiled into an overall level of complexity, which is then plotted against the anticipated level of recovery. Based on the outcome for a given portfolio, the top quartile line has been drawn (**Fig. 37**). It allows any given field to be plotted

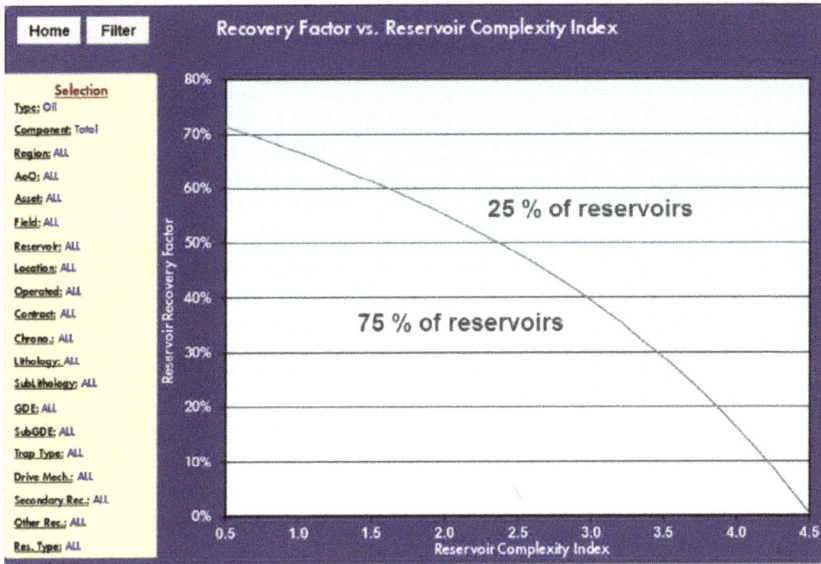

Fig. 37—Correlation between complexity and recovery (Prelicz et al. 2014).

against the portfolio. This enables an assessment of whether, given its degree of complexity, additional recovery opportunities ought to be present.

There is potential value in performing such an exercise for different parts of a field, or for different reservoir intervals within a field, to identify where it would be most appropriate to focus attention for improvement scope. It would also be beneficial for an assessment of whether the planned recovery of a new field development is sensible. Prelicz et al. (2014) suggested that this approach in Brunei's Champion Field was instrumental in lifting recovery by 5 percentage points above that expected for the original waterflood development project.

5.2 Key Performance Indicators. Key performance indicators (KPIs) are used by most assets as visualization aids to quantify the asset's performance, and these are useful to identify specific areas requiring improvement. Such indicators can be set at a high (asset) level as well as at the working level for specific issues known to be critical to field performance. KPIs can be set using specific volume expectations or they could also be based around a traffic-light system that indicates whether the issue in question is performing within acceptable norms.

Six factors to be monitored by KPIs have been reported for waterfloods in Malaysia, in combination with a number of component KPIs within each main area (Faiz et al. 2019):

- Water-injection module efficiency, supported by measures for design capacity usage, treatment capacity usage, treated water efficiency, and on-spec water usage
- Uptime and operational efficiency, supported by individual KPIs for injection module availability, injection module reliability, and injection-well uptime

- Surveillance compliance
- Subsurface conformance, supported by four sub-KPI measures on pressure conformance, injection compliance, production conformance, and cumulative voidage replacement
- Cost optimization (unit injection cost)
- Production attainment (production/injection ratio)

The KPIs to be used for any given application will depend on which factors are most critical to overall field performance, so they will need to be built on a case-by-case basis. The manner in which the KPIs are monitored might depend on the type of parameter. Factors such as simple injection volumetrics could be monitored by gauging what percentage of the target is met while more complex factors such as water quality might best be monitored by a traffic-light tool based on a number of different input monitors.

5.3 Water Shutoff. The volumetric sweep in a waterflood can vary significantly. When a field performs more poorly than was anticipated at the outset, it is usually because of poor volumetric sweep inducing high proportions of water cycling and leaving large volumes of unswept oil. This implies a high target volume for flood remediation. Water shutoff is a very important means to achieve this.

Profile modification to change the flow pathways of water through the reservoir can be performed in either the injection or the production wells. Treatments at injectors are generally preferred because the treatment is then performed at the source. However, there might be cases when this is not possible, such as when induced fractures connect different reservoir units, for example. Treatments might be performed because a zone is already swept or is taking a higher proportion of the injection volume than demanded by the reservoir management policy. Applying water shutoff in this latter case needs careful consideration because more than one pore volume of water is often needed to achieve adequate sweep, so it is possible that producible oil, as well as water, are subsequently isolated.

Data collection to understand the nature of the problem is necessary for production profile modification to be successful. This should be linked to a sound understanding of the reservoir architecture to decide which technologies might or might not offer a lasting solution to the problem. Some type of reservoir barrier usually needs to be present for a water-shutoff treatment to be effective. This is because, where such barriers are absent, a wellbore water-shutoff treatment might not be effective because the water is already very close to the well and it is likely to find an alternative route to the well. As a result, any benefits from such treatments are unlikely to be very long lasting.

In some cases, there will be functionality already installed within the completion that enables water shutoff to be performed at little or no cost. For example, where inflow control valves (ICVs) are installed, water can be shut off quickly, which has the benefit of being reversible if the change does not convey benefits. However, such options have significant initial costs, and in many cases, they will not be economically justified. In those cases, there is a range of more permanent shutoff options, which will be the focus of this section.

Mechanical options are generally the most reliable remedial shutoff options, but there are cases where chemical options might also be appropriate. The circumstances in which each option might be used are therefore also discussed.

When considering water shutoffs, problem diagnosis is an important first step. This is because, when water breakthrough in producing wells first occurs, the flow pathways associated with that water are unlikely to have swept all the available moveable oil. Thus, when considering shutoff, it is important to first consider the concept of good water or bad water (Kabir et al. 1999). Good water continues to sweep oil toward the producers. Bad water offers nothing in terms of sweep and simply follows watered-out pathways through the reservoir. The latter water is the main target for water-shutoff treatments.

Causes for bad water can include

- Reservoir channeling
- High-permeability streaks
- Fractures, faults, or other macrogeological features
- Excessive induced fracture growth
- Small, high-permeability thief zones
- Fingering
- Flow behind pipe

Relative permeability effects mean that oil will continue to be produced long after water breakthrough has occurred in heavy-oil systems. Thus, the identification of candidates for shutoff in such systems is more challenging and might only be possible as the flood becomes quite mature.

It is therefore evident that problem diagnosis, in which the location and type of influx are identified, is an important first step for successful water-shutoff treatment.

5.3.1 Mechanical Shutoff. In cases where water influx occurs at the toe of a producing well, water can be mechanically shut off by setting a bridge plug above the water-influx point. Through-tubing-deployed inflatable bridge plugs have been used to shut off water in both openhole and casedhole completions in carbonate reservoirs in Saudi Arabia (Mohammed et al. 1998). Success has been reported for both scenarios, provided that a gauge hole is present for the openhole completions to enable an effective reservoir seal. The costs for such treatments are modest in comparison with full workover costs. A vertical reservoir barrier is needed between the water-influx point and the subsequent producing interval to ensure success. More than 90% of the 40 treatments reported in Mohammed et al. (1998) were successful, adding a total of 88,000 BOPD and reducing the produced water rate by 34,000 BWPD.

A review of the success of a large number of bridge plug applications in Egypt found that the required setting depth and reservoir conditions should be evaluated before any treatment decision is made. It also found that scale removal was a necessary prerequisite for effective isolation to be achieved (A/Fotuh and Macary 2000).

This technology has been extended to horizontal openhole completions in which cement was pumped after the packer had been set to cap it and in which a gel was pumped before the cement to minimize problems related to slumping of the cement (Al-Shahrani et al. 2007). This work confirms the view that cement is often useful to further improve the seal afforded by the bridge plug.

One limitation of the bridge plug is the expansion ratio that can be achieved (although that will not be a concern if monobore completions are used). However, the capabilities in this respect are improving. For example, an application was reported where the zone to be shut off was in a 9⅝-in. casing and the deployment had to be achieved through 2⅞-in. tubing (Abdulhadi et al. 2018). Deployment by means of slickline resulted in a very cost-effective solution.

The main limitation associated with the use of bridge plugs is that all production upstream of the plug is lost. This is not a problem when the zone that needs to be isolated is at the toe of the well. However, in many cases, production from upstream of the zone that needs to be isolated must be retained. This problem can be overcome through the use of plugs and an associated straddle assembly. One such deployment using e-line is illustrated in **Fig. 38** (Rajamani and Ipsen 2016).

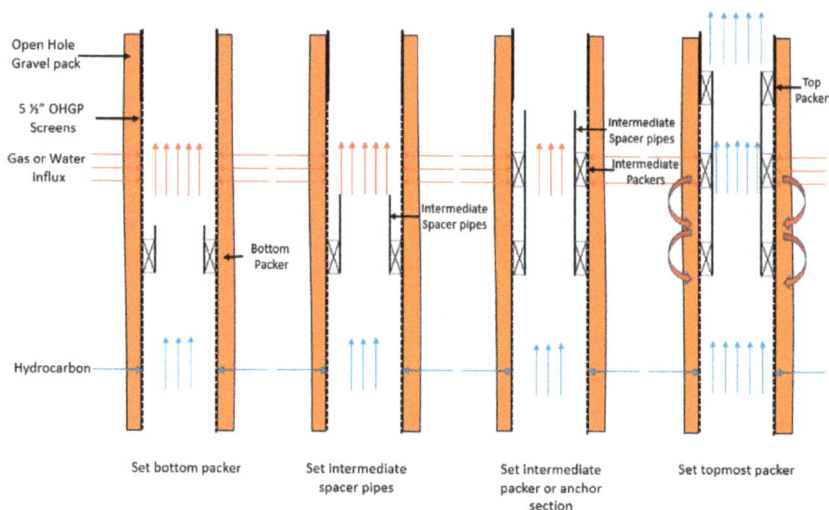

Fig. 38—Water shutoff with a straddle assembly (Rajamani and Ipsen 2016). OHGP = open-hole gravel pack.

The majority of such applications tend to be deployed in producers, but they can also add value in injection wells. One such application was performed where an injection interval representing 9% of the interval was taking 30% of the injected water (Addoun et al. 2011). This resulted in a pressurization of that interval that subsequently resulted in crossflow when the well was shut in. In this case, the wells had 7-in. monobore completions that facilitated running the straddle assembly in a single trip. The installation successfully resolved the problems. Analysis showed that, after 2½ years, there was still no leakage into the thief zone at a differential pressure of 2,600 psi.

Casing patches are another shutoff option that can provide a high degree of mechanical reliability (Forbes and Taggart 1998). Expandables provide a similar capability (Matthew et al. 2004). A clad-through-clad system removed the limitation of not being able to perform any activity below such an installation (Bourgoin et al. 2014).

5.3.2 Expandable Zonal Inflow Profiler. In cases where the installed completion enables annular flow, such as in sand control completions, the options for normal wellbore mechanical shutoff will no longer work because water flow by means of the annulus is still possible. This problem has been addressed in a number of fields in Oman through the use of so-called expandable zonal inflow profiler (EZIP) completions (Al-Mahrooqi et al. 2007).

EZIP completions use water-swellable elastomers that provide an immediate seal for unwanted features. They also enable segmentation in the annulus to facilitate future wellbore water-shutoff options. Both oil- and water-swellable elastomers are available, but in this case, water-swellable elastomers were selected. The rubber elements were vulcanized on to a standard base pipe (**Fig. 39**), with locations for the elastomer sections selected to provide an appropriate isolation capability in the event of significant water production. The elastomers were selected to expand when run in the hole based on reservoir temperature and formation-water salinity. The swelling is an osmotic-based process that normally takes place at temperatures between 50 and 90°C, and expansion is more than 100% within 1 week.

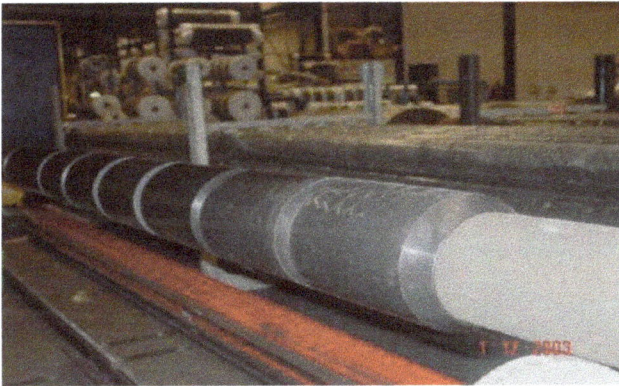

Fig. 39—Casing joint with swellable elastomer seals (Al-Mahrooqi et al. 2007).

Most of the fields in which this technology was first applied were completed open hole in horizontal wells with standalone screens. In these cases, water tended to short circuit through a natural-fracture system, so the identification of segmentation locations was determined by losses suffered under underbalanced drilling.

Fig. 40 gives an example of an initial water-shutoff treatment using this technology, where the well was separated into five segments. The first isolation segment was installed around a zone of desaturation from 962 to 1012 meters along hole (mah). Two production compartments were then selected from 1012 to 1066 mah and 1076 to 1120 mah. The last elastomer section was installed from 1194 to 1198 mah to isolate a suspected fault. The well came on stream at 88% water cut, but this soon dropped to 20% and was stable from that point.

Fig. 41 shows segmentation of a well in which a remedial EZIP treatment was applied. The producing water cut had reached 95% when action was taken, and

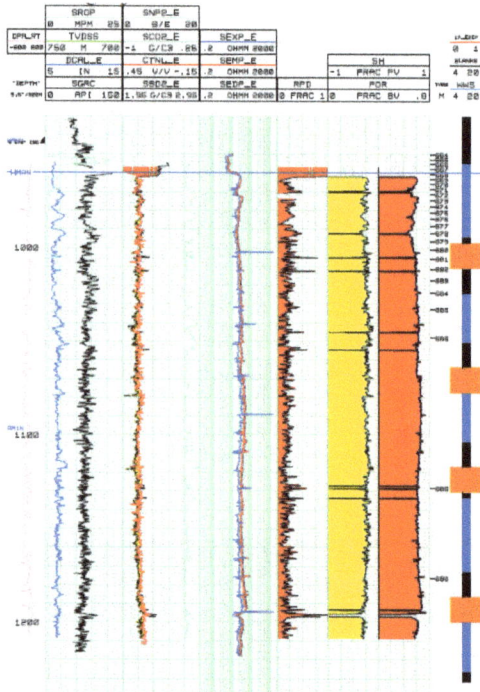

Fig. 40—Initial well segmentation using EZIP (Al-Mahrooqi et al. 2007).

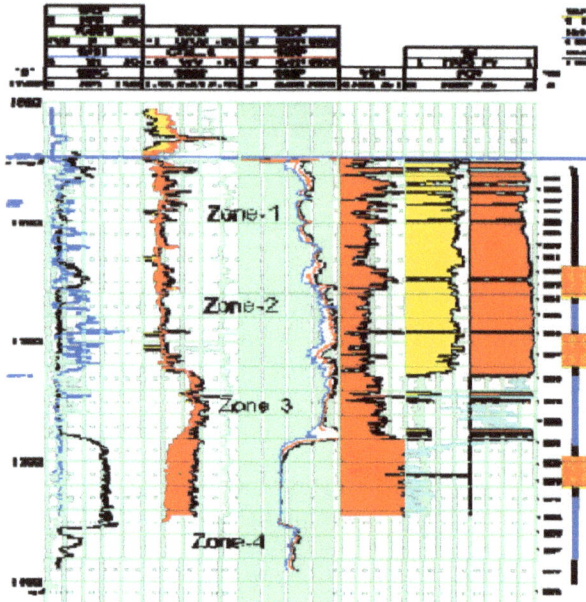

Fig. 41—Remedial action example using EZIP (Al-Mahrooqi et al. 2007).

most of the water was suspected to be coming from Segment 3. A fault was suspected to be producing the water at this location because drilling had shown an increase in the fluid gradient there. The subsequent Zone 4 was suspected of being desaturated, so the decision was made to leave only Zone 2 open to production. The well came back on stream at 50% water cut and the net oil rate increased from 7 to 60 m³/d, with the result that the well produced nearly 13 000 m³ of incremental oil.

The applications in south Oman were in sandstone reservoirs where remedial action was based on the resolution of problems caused by the open annulus in the installed sand control completions. However, the technology has also been used as a pre-emptive reservoir flow control philosophy to isolate fracture networks in carbonate waterflooded reservoirs in north Oman (Welling et al. 2007).

Both pre-emptive and remedial applications of this technology have been shown to deliver impressive benefits. A simulation of the impacts of a pre-emptive EZIP installation suggested that, in that field, isolation of 10 m on either side of a fault is needed to provide good isolation and would be expected to defer water production by up to 600 days. A remedial application in one Nimr well reduced the producing water cut from 90 to 10% (net oil rate gain of 1,000 BOPD). Although this gradually returned to 80%, an incremental 70,000 bbl oil was produced by the well, and the producing WOR was only 1.5 m³/m³ as compared to the 11.5 m³/m³ that would have been delivered without the treatment (Medeiros et al. 2004).

The only alternative to this type of technology in wells where an open annulus precludes wellbore shutoff would be to perform a chemical relative permeability modification treatment (which is discussed under Section 5.3.4).

5.3.3 Inflow Control Valve Usage. In many fields, the initial development plan specifies the installation of completions that feature an installed functionality that enables shutoff without the need to resort to external treatments or even workovers. Many wells now have ICVs installed that enable full control of flow into (in an injector) and out of (in a producer) a reservoir interval. The question is how that capability can be used to best effect. Temperature profiles can be used to identify locations of water influx and, with a model, these can be used to equalize the production profile through ICV activation (Li and Zhu 2011).

In most applications, the simulation model will be used to update understanding of reservoir flow. It will then be used to appropriately activate the ICVs to an optimized position. A workflow has been described that can be operated in either a proactive or reactive mode (Carvajal et al. 2014). The reactive mode is activated when the water cut increases and the oil cut or flowing bottomhole pressure decreases. At this point, the workflow performs a local history match to update the well performance model. It then optimizes the setting of the ICVs to minimize water production and maximize oil production for the given constraints. The proactive approach is applied every month or quarter when simulation is performed to evaluate hundreds of scenarios to obtain the best combination of ICV settings.

As the number of wells, and segments within those wells, increases, the number of potential scenarios to be managed increases dramatically. For such scenarios, an ensemble-based production optimization workflow can be used (Pajonk et al. 2011).

5.3.4 Chemical Shutoff. There are a number of different types of chemical water-shutoff options available. Each has a different application environment. The full blocking chemical options provide the same functionality as the mechanical water-shutoff technologies. The relative permeability modifiers can be selected when there is a desire to shut off water but the location of the water influx is not known. Finally, there is a group of chemicals designed to perform water shutoff in situ, deep within the reservoir. Each of these options will be discussed in turn.

Full Blocking Chemical Shutoff. Gel treatments have long been available for water shutoff when the location of the water influx to be isolated is known. Most of the chemical systems are composed of a synthetic polymer or biopolymer and a cross-linker, which form a gel in situ, although other systems based on inorganic polymers and resins have also been used. A range of different options is available from the various service companies.

In pumping such systems, some type of system must be in place to ensure the chemical is located across from the interval that needs to be isolated, rather than being injected into intervals that are still needed for production or injection. This requirement could mean the chances of success are somewhat lower if the treatment is bullheaded rather than placed. One option is to use inflatable packers to ensure the gel treatment is placed exactly where required (Plante and Mackenzie 2000).

The impacts of permeability also need to be considered. In a low-permeability interval, there might be problems achieving adequate injectivity during pumping, although, in general, it might be rare to see cases where water shutoff is needed for a low-permeability interval in a waterflood setting. In high-permeability zones (which is a setting where water shutoff is commonly needed), there can be problems associated with the high drawdowns experienced in such intervals, which might prevent the gel block from holding up in the longer term. A higher polymer concentration is a potential solution to that problem in principle, although the exact criteria linking polymer concentration to drawdown appear to be somewhat ill-defined.

In addition to ensuring that gel treatments are properly placed, proper consideration must be given to the time needed for gelation to occur given the reservoir conditions (especially temperature). Based on the findings from 12 relatively early treatments, three key guiding principles have been identified (Fulleylove et al. 1996):

- Make sure the treatment is safe to deploy and will not cause short- or long-term environmental damage. These treatments were based on a chromium crosslinked polymer system.
- Understand where the oil production will come from after the treatment. Make sure the treatment does not damage that potential by ensuring not only that the chemical is put in the right place, but that there will be no indirect damage resulting from water coning or crossflow.
- Ensure the gelant will go into all target zones to the required depth and that it will remain fully active until it gels.

In a review of more than 300 gel polymer treatments, it was suggested, perhaps optimistically, that the success rate for these treatments in new areas is 75%, rising to 95% or more in areas where there is experience with the technology (Portwood 1999). Such numbers do appear to indicate success for this technology, but there is no clear understanding of the volume of chemical needed to deliver an effective treatment, and there is some general acknowledgment that treatment sizing is subjective. Given this difficulty, it is difficult to see how such a high treatment efficiency is achieved.

In general, the deeper a treatment is placed (i.e., the more that is spent on treatment volume), the longer it is likely to be effective because it should take longer for the water to find an alternative route back to the wellbore. This required treatment depth is a significant impediment to successful treatments because operators are loathe to spend money on a chemical that is not needed, so one reason for failure could be that insufficient chemicals are pumped. Treatment economics should be taken into account, which places downward pressure on pumped volumes. However, economics are never favored if the subsequent treatment is ineffective.

Several rules of thumb for treatment volumes have been proposed. One suggests placement of gel to a depth of 50–60 ft from the wellbore, and another recommends a pumping volume of 50–200 bbl of gel per perforated foot. The poor understanding of exactly what volume is needed for effective treatment is one reason poor treatment success is sometimes experienced, especially in fields where such treatments have not been pumped before.

The required gel strength represents another element of uncertainty. There is an accepted notion that increased gel strength is needed for zones producing greater volumes of water, and perhaps also where a higher pressure differential will need to be held by the plug. However, it is difficult to be more specific in defining the required gel strength.

Candidate selection is another critical aspect for the successful deployment of chemical shutoff treatments (Kabir et al. 1999).

In some cases, it might be appropriate to combine gel treatments with cement (van Eijden et al. 2004). This gel/cement system combines the properties of two different shutoff techniques: cement for mechanically strong perforation shutoff and gel for isolation of the matrix. This system also appears well-suited to applications where high pressure differentials have to be tolerated.

The shutoff of natural fractures in carbonate reservoirs is an area where gelled polymer treatments have been used but where an even greater level of uncertainty is observed regarding the required treatment design. Indeed, field experience suggests that results in the treatment of this problem are wide ranging, and although some applications have been very successful, little or no benefits have been observed in many treatments. This might not be particularly surprising because different fields can have different fracture widths, fracture conductivities, matrix permeabilities, wettabilities, and relative permeabilities. Thus, these factors might be expected to result in variable responses to such treatments.

Conformance improvements in the Grayburg Zone, in West Texas Field, have been performed with gel treatments, and treatment sizing has been based on the outcome of tracer studies (Shook et al. 2009). However, subsequent gel volumes were significantly less (approximately < 50%) than those that had been used for historical gel jobs, which has been found to be detrimental to the economic success of treatments (Jain et al. 2020).

In one treatment pumped in an offshore Californian well, the treatment shut off 50% of the water production and increased the oil production by 300%. However, the benefits only lasted for 3 months (Cheung and Quilter 1998). In contrast, five treatments at a field in southeast Turkey were more successful. The benefits resulted in a payout of the investment within 60 days and a much more extensive production improvement period (Canbolat and Parlaktuna 2012).

There are a number of factors that could potentially influence the success of treatments in this setting, and there is a poor grasp of which factors might be the most important to any given application. Thus, it appears that the success criteria for each field might need to be developed using a hit-or-miss approach until what works in any given setting is properly established. The required depth of treatment will be a critically important consideration to define if success is to be achieved.

Chemical Relative Permeability Modification. In cases where it has not been possible to identify the location of water influx but there is still a critical requirement to reduce the water production, the options that have been discussed can no longer be applied. The only option available is to pump a chemical relative permeability modifier treatment.

These chemical systems target improvement of the profile in producing wells. They are designed such that, when the chemical is pumped, the water production from the well will be reduced significantly while the oil-producing intervals will be affected to a smaller extent. At a given tubinghead pressure, the reduction in water production will result in a better lift efficiency. Consequently, the oil production capability of the well will be expected to improve. Although the oil-producing parts of the well are less affected than the water-producing intervals, they are still affected to some extent. It is therefore inevitable that a relative permeability modification treatment results in a significant increase in skin.

For the application of a self-selective system, it is vitally important to understand the reservoir conditions under which the treatment will be carried out. These must be conducive to the chemical treatment. Otherwise, it will fail even though the chemical system might have been proven to work in a laboratory environment.

For a completely homogeneous reservoir, an effective self-selective system will reduce both the water flow and the oil flow. It therefore acts as a choke, and both the water rate and the oil rate will be reduced to the same degree because the chemical system causes water holdup in the near-wellbore region. While the flow of water into the wellbore is reduced, the water flows deeper within the reservoir will remain unchanged. Consequently, the water saturation in the treated area will increase. The water relative permeability will also increase after having first been reduced by the chemical system. After some time, the ratio of the relative permeability to water to the relative permeability to oil will become identical to that before the treatment, but the absolute levels will have been reduced. In other words, the treatment results in an increase in skin around the treated wellbore. Therefore, for constant drawdown conditions, the oil and water production rates have decreased, but the ratio of the water rate to the oil rate has remained unchanged.

It follows that a treatment with a bullheaded relative permeability modifier is only potentially attractive if all the following conditions are met:

- The reservoir at the location of the well to be treated has several intervals.
- The intervals have no pressure communication with each other.
- The intervals produce either mainly water or oil.

The systems that are able to reduce water mobility, but have a limited impact on oil mobility, are based on polymer retention. Over the years, vendors have promoted a number of products they claim are effective relative permeability modifiers. However, the results have not been consistently encouraging. On the basis of industry experience to date, it is probably accurate to say that the chances of success with these treatments have been lower than those experienced for full blocking technologies. As a consequence, such treatments tend to be a last resort when other options for well improvement have been tried and have failed, or in cases where no other options for shutoff are available.

Two successful field trials with one of the available technologies have been reported (Kume 2003). **Table 1** presents the results. Although there has been a significant increase in skin associated with both treatments, the reduction in water production enabled an increase in the subsequent oil production. However, these results are somewhat short term, and it is not known how persistent the gains were.

Table 1—Successful relative permeability modification treatments (Kume 2003). WHP = wellhead pressure.

	Water Cut	Total Rate (B/D)	Oil Rate (BOPD)	Water Rate (BWPD)	WHP (psi)	Skin
Case History 1						
Prejob on 72/64ths	89	897	100	797	180	97
Post-job on 26/64ths	72	385	108	277	345	322
Post-job on 72/64ths	72	714	200	514	180	322
Case History 2						
Prejob on 72/64ths	87	2,346	305	2,041	135	47
Post-job on 26/64ths	73	460	124	366	525	123
Post-job on 72/64ths	73	2,014	544	1,470	135	123

The appropriate selection factors for such treatments have been considered (Ahmed et al. 2010) and used to define criteria that would make a well either a good or a bad candidate for treatment. These are presented in **Table 2**.

Table 2—Criteria defining good and bad candidates for relative permeability modification (Ahmed et al. 2010).

Good Candidate	Bad Candidate
Multiple production zones with at least one clean oil-producing zone	Single production zone with both oil and water flowing together
No crossflow between the clean oil zone and other watered-out zones	Crossflow occurs between producing intervals
Well not fully drawn down with high productivity index	Low productivity index with no scope for increasing drawdown
Moderate values for water cut (in the range of 50–95%)	Very high (> 97%) or very low producing water cuts
Permeability contrast between oil- and water-producing intervals	No permeability contrast or lower permeability in the water-producing interval

These criteria are supported by Sydansk and Seright (2006). The crossflow criterion does not refer to the potential of crossflow from one zone to another by means of the wellbore but to crossflow between the watered-out zone and the oil-bearing interval behind any inserted block within the reservoir (**Fig. 42**).

These constraints suggest the application for such treatments might be somewhat limited. In general, it appears unlikely that horizontal wells will be good candidates

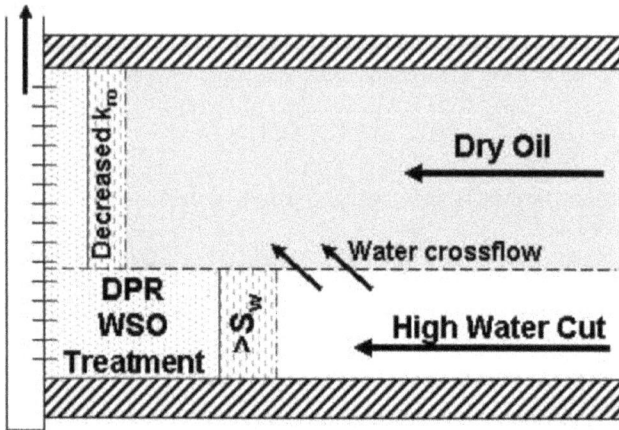

Fig. 42—Crossflow reduces the impact of relative permeability modification treatment (Sydansk and Seright 2006). DPR = disproportionate permeability reduction; WSO = water shutoff.

for treatment. Furthermore, even when conditions meet the requirements for such treatments, they might not necessarily result in economic treatments. It also appears likely that, without extensive field experience, the success ratio for such treatments will be low, although individual successes are still possible. Faber et al. (1998) suggests a success rate of approximately 40%, indicating that additional work is required to improve the understanding of these systems through a better insight into the chemistry and factors such as the required injection concentrations and volumes.

Deep In-Situ Shutoff. In cases where injection water breaks through in the producer by means of a high-permeability streak in a zone that can be isolated from the rest of the producing well by geological barriers, the standard approach is to isolate that interval using either mechanical or chemical plugs that are set at the point of ingress to the well. This works very well when the geological barriers are laterally extensive. It becomes less efficient when those barriers become less extensive because there is the potential for the water to find its way around the barrier and back to the well.

In some cases, there might not be a geological barrier. When this happens, the high-permeability streak is likely to induce flood short circuiting, and the adjacent lower-permeability intervals might not flood at all, or flood very slowly, because of the permeability contrast and relative permeability effects. In such cases, wellbore treatments are ineffective and another methodology is needed to change the flood streamlines. One option is to change the pressure sinks in the reservoir through cyclic injection (see Section 5.7), but that option is often economically viable only when the flood is quite mature.

The alternative is to perform a diversion treatment deep within the reservoir between the injection and production wells to divert the injected water from the high-permeability zone that is dominating flow into the more poorly swept zones, thereby improving oil recovery. The amount of oil that is being bypassed as a result of this problem will be a function of the degree of heterogeneity as well as the aspect ratio between the volumes in the unswept reservoir and that part of the reservoir acting as a thief zone.

The aspect ratio can be expected to be very important in establishing whether a remedial treatment is economically justified. This is illustrated in **Figs. 43 and 44.**

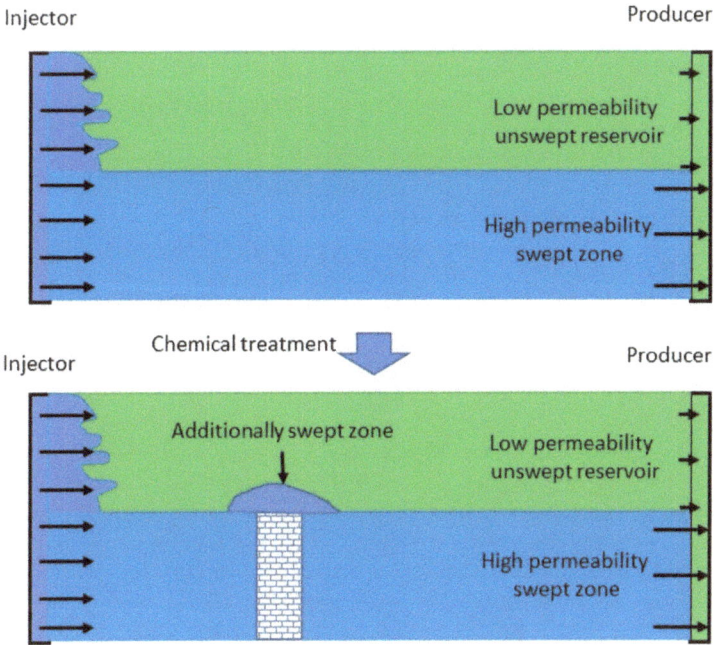

Fig. 43—Poor aspect ratio for deep in-situ water shutoff.

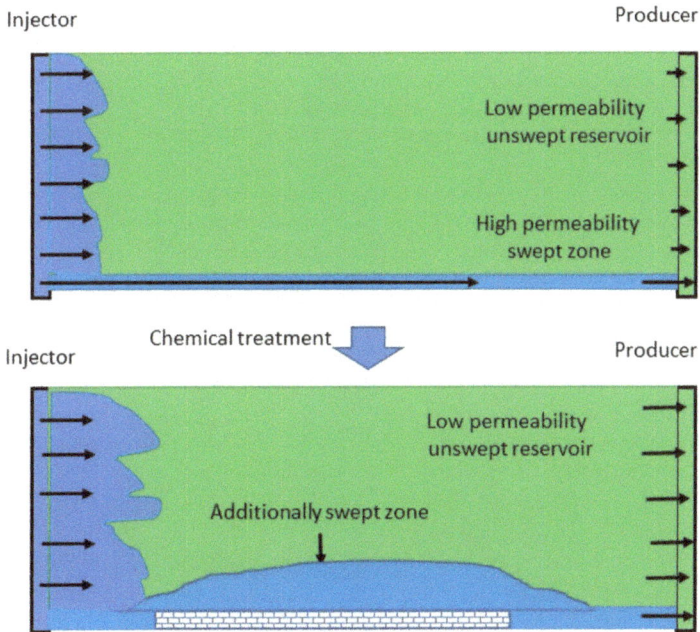

Fig. 44—Good aspect ratio for deep in-situ water shutoff.

In Fig. 43, the swept zone is a significant proportion of the total reservoir volume. As a consequence, a large volume of chemical might still only deliver a plug that covers a small proportion of the reservoir. In this case, the diverted water might only sweep a small incremental volume of oil before it finds its way back into the already-swept zone.

In contrast, in Fig. 44, there is a thief zone that is relatively small in volume. Therefore, it could be relatively easy to set a plug that diverts the water to sweep an appreciable incremental volume of oil. Seright et al. (2011) suggested that the low-permeability layer should have a thickness of at least 10 times that of the thief zone.

One of the chemicals that performs this deep in-situ shutoff is based on a polymer that can be dissolved in the injection water and has particle sizes sufficiently small that it is able to pass through the rock pores with flood water (Frampton et al. 2004). Initially, the polymer will be at the temperature of the injection water, but as it passes through the reservoir, it will gradually warm to the reservoir temperature. As the polymer heats up, it expands. This blocks the reservoir pore throats it contacts and diverts any water that subsequently follows the same pathway. The polymer particles have an initial average diameter of 200–300 nm and expand by a factor of 4–10, with the exact degree of expansion dependent on the salinity of the injection water. (A much larger expansion factor is seen with low formation-water salinities.)

It follows that simulation studies are needed to determine the optimal slug size and activation time of the polymer for any given reservoir system. The expansion time at a given temperature is controlled by the number of reversible crosslinks between polymers. Consequently, different grades of this system are available, so the activation time can be tailored to the application. Low levels of permanent crosslinks prevent the polymer from completely unraveling.

The polymer has a second diversion mechanism. The swollen polymer can further adsorb and reduce water flow in a manner similar to the resistance factor observed in polymer floods with hydrolyzed polyacrylamide.

Based on the mechanism, this technology might find application in cases where

- High-permeability streaks or thief zones are negatively impacting vertical sweep
- Improved sweep of channel margins is required
- Architectures have direct communication between high- and low-permeability intervals and those intervals exhibit a high-permeability contrast

This system was first deployed in Indonesia's Minas Field in 2001 (Pritchett et al. 2003). Incremental oil was not expected in this early application because of the flood maturity, but some was observed. This, together with the unequivocal demonstration of a block 38 m from the injector, provided the impetus for further field applications.

The first commercial applications subsequently took place at Alaska's Milne Point Field in 2004 (Ohms et al. 2010). The plugs were clearly placed deeper within the reservoir compared to the earlier Minas application, and commercial incremental volumes of oil were said to have been delivered. The production response to the treatment is shown in **Fig. 45**. Subsequent successful treatments at a field in Argentina's San Jorge Basin (Mustoni et al. 2010) led to a much wider application of this technology.

MPB-04 Incremental Oil Associated with Treatment

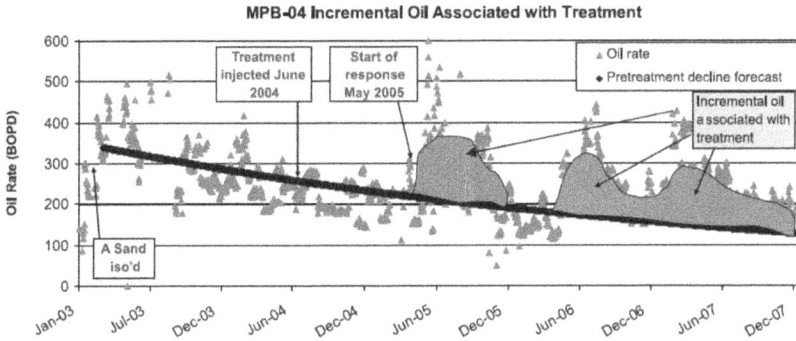

Fig. 45—Production response to in-situ polymer treatment (Ohms et al. 2010).

Analysis of the characteristics of the treated wells suggested the following application window:

- All wells cased, cemented, and perforated.
- Porosity of thief zones in the range of 19–26% and with permeability up to approximately 1,400 md. There might be an upper range of the thief zone permeability that can be successfully treated because an attempt to treat an 8-darcy thief zone (Roussennac and Toschi 2010) appears to have been unsuccessful. Although some reservoir effects were noted, no positive production impacts were observed in the offset producer.
- Thief zone thickness of up to 75 ft treated and with thief-to-pay ratios of up to 50%.
- Producing water cuts in the range of 66–99% but with the majority having water cuts greater than 85%.
- Bulk permeability to thief zone permeability contrast in the range of 2–7.
- Sandstone reservoirs; this technology is not currently applicable to carbonate reservoirs.

Using these types of criteria, treatment successes in the range of 80% have been obtained (Mustoni et al. 2010).

When such treatments are considered, experiments need to be performed to assess whether the polymer fulfills the requirements of injectivity, activation time, and a permeability decrease of the high-permeability streak. Such experiments form the basis of the treatment design.

For subsequent field applications, very simple equipment is required: metering equipment for the polymer and the surfactant connected to the water-injection line. Co-injection of a surfactant inverts the polymer emulsion and facilitates good dissolution of the polymer in the injection water.

This technology has been compared to conventional polymer flooding. Seright et al. (2011) suggest that while short-term economics tend to favor in-situ profile modification technology, it does tend to deliver incremental volumes that will be much lower than those achievable from conventional polymer flooding. Also, in the longer term, this latter option might be better both in terms of recovery and economics. However, this should be assessed on a case-by-case basis, and individual circumstances could differ in different fields.

An alternative that essentially performs the same function is to inject a conventional polymer followed by a multivalent metal ion solution as a crosslinking agent (Manrique et al. 2014). This option is referred to as in-depth colloidal dispersion gel (CDG) technology. It was first deployed in the North Burbank Unit in Oklahoma (Moffitt et al. 1993). However, the prime motive in that case was to reduce the volume of polymer needed for flooding rather than to achieve in-situ diversion.

CDG technology has been deployed in the US, Argentina, Colombia, and China with either aluminum citrate or chromium acetate as the crosslinking agent. These treatments appear to be applicable to a reasonably wide range of reservoir conditions, and treatment conditions have also varied:

- Temperature: 80–210°F
- Permeability: 10–4,200 md
- Oil viscosity: 5–30 cp
- Polymer concentration: 250–1,200 ppm
- Polymer-to-crosslinker ratio: 20:1 to 80:1
- Volume injected: 10,000–650,000 bbl

This option is used primarily for in-situ diversion rather than as a method for influencing polymer flooding. This means that pumping volumes are typically at the lower rather than the higher end of this treatment volume range. Treatments show that there is no significant injectivity problem as a result of such applications. However, as **Fig. 46** shows, a comparison of Hall plots for thermally activated polymer treatments and CDG treatments of a similar size indicates that the CDG treatments exhibit increased impairment-type behavior. This is not particularly surprising because the thermally activated polymer might not see an increase in viscosity so early and could imply the achievement of a deeper placement for the thermally activated polymer treatments in comparison to CDG treatments.

Another option is to create a gel at surface conditions (Delshad et al. 2013). This option might overcome any problems from uncertainties associated with control on gelation time or with any adsorption or shear degradation that could occur. However, despite these potential benefits, uptake of this technology does not appear to have been as widespread as for the other technologies described.

Another chemical that has been used for in-situ shutoff is the injection of sodium silicate. This chemical has a pH in the range of 11–13, with the pH depending on the exact composition and the molar ratio, n, in the formula $(SiO_2)_n \cdot Na_2O$. At high pH, the dimer silicate species predominates. As pH is reduced, polymerization of the silicate species begins, thereby initiating a gelation process. The minimum gelation time occurs just below pH 7. Work has been conducted to define the conditions required for such an application in Norway's Snorre Field (Stavland et al. 2011). Dilute hydrochloric acid was used as a gel activator, but it was found that, because silicate precipitation occurred when the treatment mixed with formation water, water needed to be preflushed before pumping the treatment.

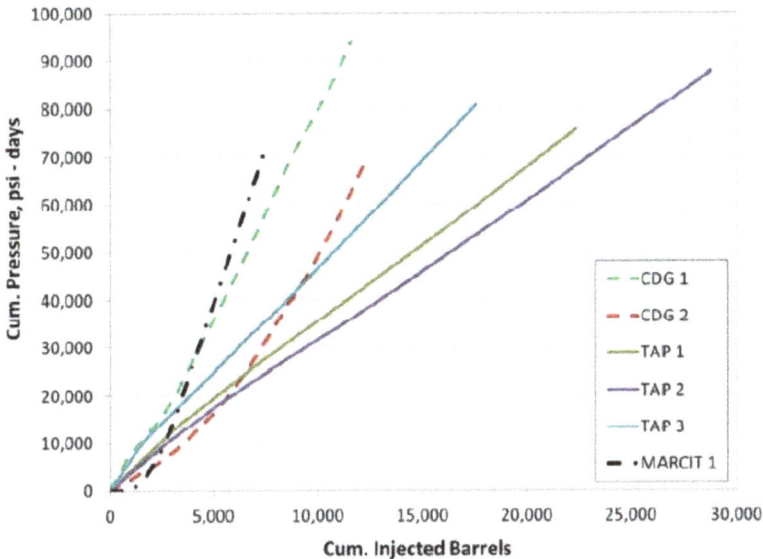

Fig. 46—Hall plots for CDG and thermally activated polymer (TAP) treatments (Manrique et al. 2014).

Indeed, premature gel formation is probably the major risk for this technology, but a field pilot was subsequently deployed (Skrettingland et al. 2014).

5.4 Stimulation/Perforation. Over time, it is possible that damage in the near-wellbore region of both production and injection wells will induce well impairment and also influence the flow pathways such that the reservoir sweep will be negatively impacted. In such cases, stimulation can both improve well productivity and remediate sweep.

Of course, when stimulation is attempted, it is important to understand the nature of the damage so an appropriate stimulation treatment will be delivered.

A good example is a laminated fluvial system where injection conformance was a problem (Kothiyal et al. 2012). An attempt to restore adequate injection distribution through reperforation of the damaged interval was unsuccessful, and it was subsequently found that wax dropout and poor injection-water quality were responsible for the issues. As a result, wellbore heating and a surfactant treatment followed by acidization were an effective well remediation. A straddle assembly was useful in targeting the stimulation location, and this was very successful in improving flood conformance.

Therefore, in stimulations in waterflooded areas, placement should be strongly emphasized to ensure that sweep does not become compromised as a result of the creation of preferential flow pathways. A number of different placement and diversion techniques are available, including mechanical methods and chemical diversion.

The mechanical methods—which include the use of packers, coiled tubing, and ball sealers—are usually the most reliable means of achieving diversion, although they do increase the cost of treatment. Furthermore, these options cannot be applied in gravel-packed wells where the completion does not allow mechanical isolation at the wellbore.

In such cases, there is no real alternative to the use of chemical diverters. These are normally materials that are insoluble in acid but highly soluble in the injection or production fluids (as appropriate) that can be used to form a thin, low-permeability filter cake at the formation face. Having acted as diverters, they can then enable a rapid and complete cleanup. The advantage of chemical diversion is that it does not require zonal isolation to work. The main concern is that, if the diverter is not removed by the produced fluids, it can also cause impairment. Alternatively, diversion can be achieved through the use of particular fluid types. Foam, viscous fluids, gelled acid, emulsified acid, and crosslinked acid have all been used.

Preferential stimulation of watered-out zones must also be avoided. This is a particular risk because watered-out zones are likely to have somewhat higher permeability than unswept zones and so will be naturally prone to accepting most of the injected stimulation fluid. In some circumstances, a very well-placed stimulation treatment could help improve sweep by improving the vertical sweep profile, such as in cases where impairment has been more severe in tighter intervals than in more permeable intervals. For one case in water injectors, stimulation using chemical diversion increased vertical injection coverage of the reservoir from 22.4 to 60.9%, while stimulation with mechanical tools improved the coverage from 26 to 85.4% (Lerma and Giuliani 1994).

These considerations limit the applicability of bullheaded stimulation treatments in waterflooded areas and require that stimulation always use some form of diversion. An example where targeted stimulation was important was in the Elk Hills Field in California (Walker et al. 2002). In this case, the waterflood targeted the low-permeability tight turbidite sands. Fracture stimulation of producing wells was needed without stimulating adjacent high-permeability, water-bearing sands, so containment of fracture height in the stimulation became vitally important.

Indeed, stimulation of low-permeability waterflooded reservoirs is a double-edged sword because increased throughput can offer material recovery benefits, but these can only be realized if the treatment stimulates oil rather than water. These issues were clearly observed in the fracture stimulation of low-permeability waterfloods in western Siberia (Guglielmo et al. 2006). Many of the treatments delivered incremental oil volumes that were lower than those forecast because stimulation was based on outdated log interpretations. When water and oil production was matched in a layered simulator, a much improved stimulation performance was observed because much better stimulation of oil rather than water was achieved.

Diversion is particularly important when stimulation is needed in wells producing at very high water cuts. This is often the case in mature-waterflood producers in carbonate reservoirs. Diversion is always a challenge in heterogeneous carbonates, but it becomes even more challenging in high-water-cut environments, where it becomes crucial to stimulate the remaining oil-bearing intervals rather than the dominant watered-out zones. In a case reported from offshore Dubai, UAE, efficient diversion with an acidization treatment was achieved using a viscoelastic surfactant that maintained its viscosity in water but was easily broken down in oil (Shnaib et al. 2009). A campaign with this system was able to increase oil production by 53% while reducing the producing water cut from 83 to 79%.

A similar problem was faced in a mature-waterflooded sandstone reservoir in Colombia. The producing water cut exceeded 90% and yet a calcium carbonate scaling problem still required regular well stimulation to maintain productivity (Jaramillo et al. 2010). Straddle packers were initially used to achieve diversion across the laminated sands, but at high water cuts the costs became prohibitive. In this case, diversion was achieved using a viscous disproportionate permeability modifier (VDPM) pumped with the acidization treatment. This treatment reduced costs by 70%, and benefits (higher oil rate, lower water rate) occurred in two-thirds of well treatments. This was in marked contrast to treatments without diversion, where increased oil production was always more than offset by increases in water production. An example of the type of success from such treatments is shown in **Fig. 47.**

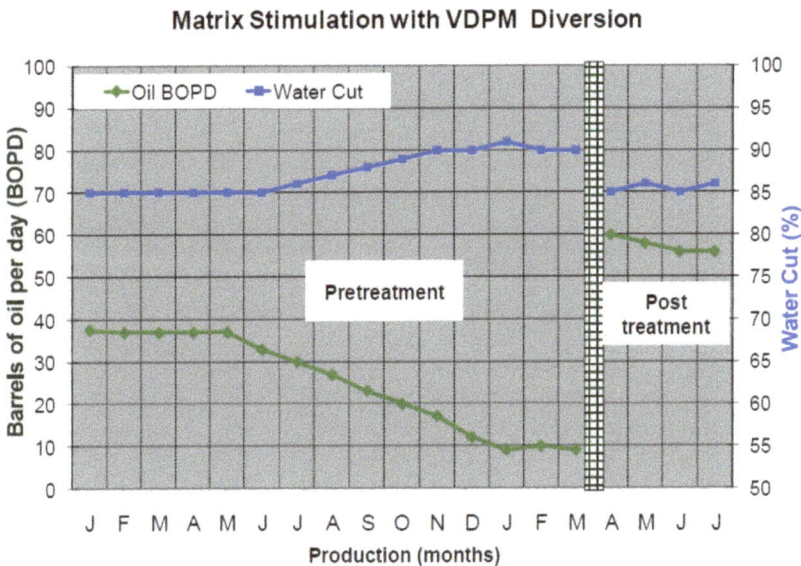

Fig. 47—Stimulation success in a high-water-cut producer (Jaramillo et al. 2010).

A general level of care is needed when considering stimulation in waterflooded carbonate reservoirs. When acidization is performed in this environment, the acid etches preferred pathways in the rock, so it tends to follow local high-permeability streaks rather than stimulating on a uniform front. There might therefore be a tendency to locally create wormholes. While this clearly stimulates well performance, such features must not connect with natural-fracture networks that are not initially connected to the wellbore, which would significantly affect the sweep. This implies that a reasonably sound geological understanding must be in place before considering well stimulation in waterflooded carbonate systems.

In waterflood systems, impairment in producing wells is often attributable to scale deposition in the near-wellbore region. Barium sulfate and strontium sulfate scales are particularly problematic because they are very hard and have very low solubilities, so they can be very difficult to remove. Provided the reservoir temperature is high enough, chelating agents can be very effective agents for the removal of such scales. At low reservoir temperatures, it is usually necessary to revert to mechanical methods such as milling or high-pressure jetting operations.

Commercial scale dissolvers are usually based on either ethylenediaminetetraacetic acid or diethylenetriaminepentaacetic acid. Despite success with these treatments, the chelation packages have the potential to cause mineral dissolution and formation damage (Jordan et al. 1998).

Stimulation fluids invariably consist of a number of different chemicals, with each designed to perform a different function in the treatment. Often, the additives are expensive, and there should therefore be a quantified need for each one used. Many additives have some surfactancy effect and can therefore influence formation wettability. Therefore, the benefits and drawbacks should be closely monitored when stimulating in a waterflood environment.

The use of surfactants in stimulation packages can be a particular concern for injection-well treatments. However, specifically selected surfactant stimulations were performed to increase water-injection rates for injection wells located in the oil column in the Magnus Field, in the UK North Sea, by improving the near-wellbore relative permeability to water (Dymond and Spurr 1988).

In water-injection wells, bacterial activity in the near-wellbore region of injectors can also be a source of injectivity decline, especially if the topside bacterial control treatment has been inadequate. In such cases, sodium hypochlorite could be an appropriate stimulation fluid (Clementz et al. 1982).

5.5 Facilities Debottlenecking and Optimization. It is expected that some degree of facilities change could be required as a waterflood proceeds because it is rare that waterflood projects proceed entirely as originally planned. This occurs most commonly because the geological complexity has been underestimated. There could therefore be consequences on the production side because the volumes of water produced might be greater than anticipated. As a result, it is possible the gross volume handling capacity, and the water handling capacity in particular, could become constrained.

In some cases, small operational changes might be adequate to deliver the required improvements. For example, when increased water volumes make it difficult to meet the water quality specifications for injection or disposal, it might be possible to make changes to the separation process. Increasing the interface levels in the separator will increase the residence time available in the water phase and could enable the required specifications to be met.

Another option for improvement within the existing infrastructure is to perform computational fluid dynamics (CFD) to establish the amount of the capacity within separators that is actually being used. If this exercise identifies significant dead spots within the vessels, it might be possible to perform remediation by making changes to the vessel internals, which will subsequently improve performance.

One such assessment was performed in an offshore facility where both the inlet and the low-pressure separator were bottlenecked (Lee et al. 2009). The CFD study of the low-pressure separator (**Fig. 48**), illustrating the flow pathways colored by velocity magnitude for the flow of oil (Fig. 48a) and water (Fig. 48b), showed that the liquid phase capacity was only 68% used. Consequently, a number of design improvements were recommended, including changing the inlet spreader, changing the perforated plate baffles, changing the weir height to operate in spillover mode, and improving volume usage and the adjustment of the separator levels to equalize the flow velocities of oil and water. These changes were expected to improve the volume usage to 94% (**Fig. 49**).

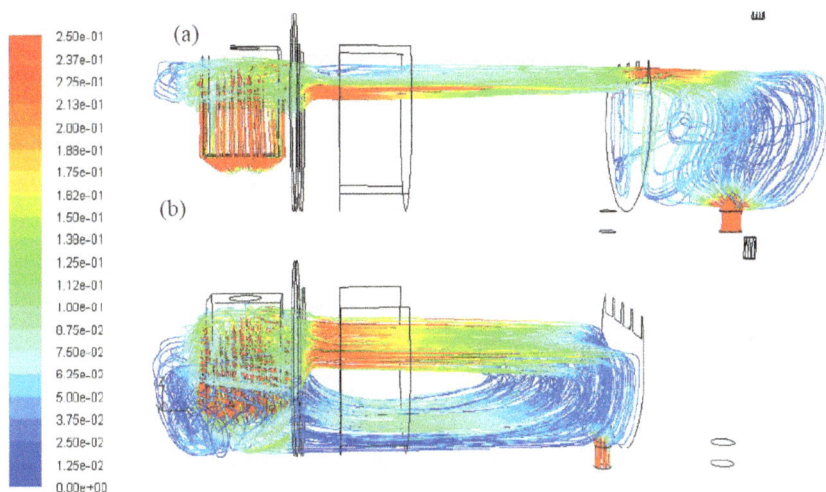

Fig. 48—Flow patterns for oil and water in a low-pressure separator (Lee et al. 2009).

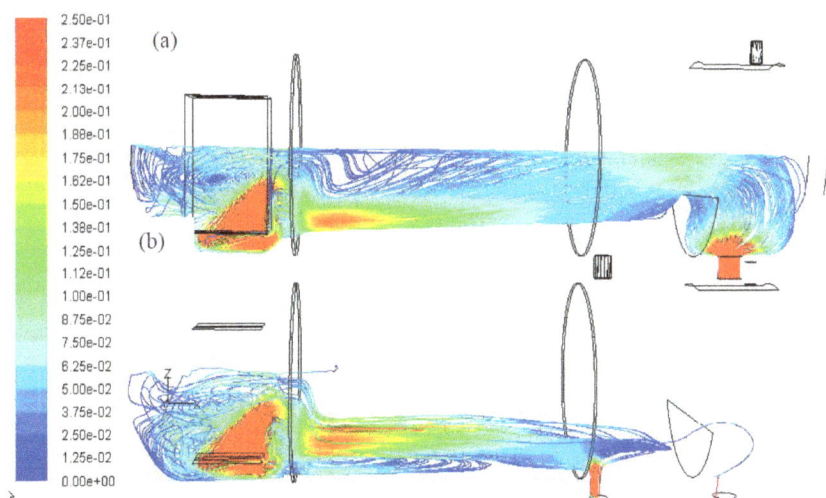

Fig. 49—Flow pattern in a low-pressure separator after design changes (Lee et al. 2009).

Another option for debottlenecking is to install additional separators. For example, it could be appropriate to install a free-water knockout tank upstream of existing dehydration vessels to reduce water loading into the separators. This option is much more likely to be deployed in onshore settings, where there is likely to be plenty of space available, rather than on an offshore platform, where there is rarely excess space available for the installation of new vessels.

A similar option is to install an in-line separator (Abbad et al. 2015). This option relies on the producing water cut being high enough for the water phase to be continuous and for the bulk of the water to have separated in the line before reaching the processing facilities, which enables the bulk water volume to be tapped off from the bottom (**Fig. 50**). The success of this option also depends on the flow regime, but if the in-line separator is installed in an appropriate location in the process, it can be an effective debottlenecking option. An in-line phase splitter, which is a compact pipe-based gas/liquid separator, has been considered for application to the Bonga floating production, storage, and offloading vessel, offshore Nigeria (Krebs et al. 2016).

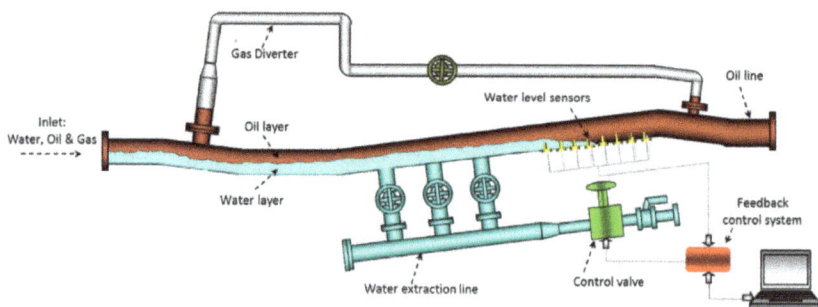

Fig. 50—In-line separator (Abbad et al. 2015).

A similar option is to install a partial separation system at satellite facilities where the bulk of the water is immediately separated and injected, with the remainder of the fluid (lower water cut oil) being routed back to the central processing facility (Rawlins 2017).

It is also possible that the injection capacity could become constrained. Often, this occurs because injection wells or infill developments are added some time after the initial development. It could also occur because satellite developments are brought onstream. The water-injection capacity at Statfjord, in the Norwegian North Sea, was expanded from an initial 736,000 BWPD to more than 1 million BWPD at high availability (Hancock 1988). In another case in Saudi Arabia, the injection plant capacity was adequate, but the injection distribution system was bottlenecked (Bangkong et al. 2017). The constraints in the system were identified through an injection system capacity test. Network pipeline modeling was subsequently used to identify the optimal debottlenecking methodology.

Routine maximization of the availability of the injection capacity is an important element of waterflood optimization. Operators routinely monitor the types of production deferment as a means to improve long-term production performance. However, there is not always the same level of focus on the injection system. It is always a good idea to track the levels of injection downtime and to categorize the types of problems that contribute to such downtime so there can be a focused effort to improve performance in the longer term. In the context of a waterflood, such efforts can be expected to result in an improvement to production performance in exactly the same manner as would occur with a reduction of production deferment.

5.6 H₂S Control and Remediation. The H_2S levels in produced fluids are measured to demonstrate that reservoir-souring control methodologies are performing as per expectations. Trends in H_2S can be used to modify the injection programs, but H_2S trends need to be closely monitored because they relate to the limits set by the installed well and topside equipment metallurgies.

Historically, many high-H_2S-producing wells have been shut in because of materials concerns. Acceptable H_2S limits for nonsour service materials are based on the partial pressure of H_2S, and they are commonly set at a partial pressure of 0.05 psi. If a well is shut in, the pressure rises, and consequently, the H_2S partial pressure will also rise, even though the amount of H_2S will not have changed. This implies that H_2S risks have been increased by the shut-in. However, H_2S concentrations are expected to continue to rise if production is continued. This suggests that wells would actually need to be shut in before the H_2S limits are reached. A mitigating factor might be that the generic H_2S limits were originally developed on the basis of conditions in wet gas wells, and the conditions in waterflood production wells might be somewhat lower risk in practice. When a dangerous H_2S condition is approaching, it might therefore be appropriate to perform corrosion testing to more accurately define appropriate limits.

Where wells are producing at dangerously high H_2S levels, it might be possible to apply H_2S scavenger by means of a squeeze treatment.

There could a number of reasons that an operational reaction to produced H_2S is needed. Those reasons can include

- H_2S limits in gas sales contracts become compromised.
- HSE exposure of staff.
- H_2S limits for installed metallurgies are compromised.

Reservoir-souring problems have induced significant amounts of H_2S production in many fields. As the level of H_2S rises, there will inevitably be some degree of H_2S treatment and removal required, which can be achieved by adding H_2S-removal process equipment. Options could include the installation of a redox process in which the H_2S is directly oxidized to elemental sulfur while ferric ions are reduced (**Fig. 51**).

The creation of elemental sulfur can induce operational problems. The original units installed by one operator typically required plant shutdowns every 6 weeks to remove solid sulfur accumulation and skid piping (Roberts and Roberts 2005). This presumably resulted in appreciable deferment (which was understood to have occurred with an installation at the Lekhwair waterflood in Oman). However, an online cleaning system improved system operation so that the units were able to perform for more than 1 year without cleaning being necessary.

An alternative is to use an amine gas-treating unit in which a downflowing amine solution contacts the sour gas to generate a sweetened gas stream. The resulting H_2S-rich amine is then routed into a regenerator to produce amine that is recycled for reuse. The stripped overhead gas then contains the separated H_2S.

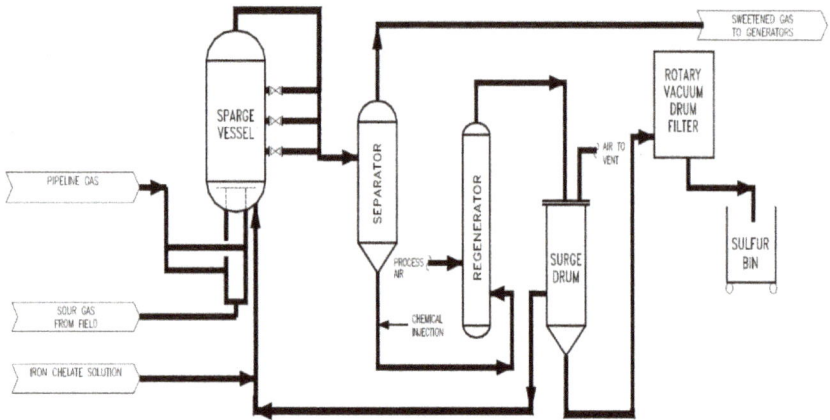

Fig. 51—Redox process (Roberts and Roberts 2005).

The primary difficulty associated with H_2S-removal process equipment is accommodating such equipment in an existing facility, where space is usually limited. Consequently, many facilities that have become sour have treated H_2S using chemical H_2S scavengers, which are usually based on triazine chemistry. In triazines, the carbon atoms of a benzene ring are replaced by one or more nitrogen atoms and, in the reaction of triazines with H_2S, the nitrogen atoms in the ring are replaced by sulfur atoms. The mechanisms are not completely understood and there are a number of reaction products, but two sulfur-containing products, thiadiazine and dithiazine, predominate.

The two most commonly used triazines for scavenging are hexahydro-1,3,5-tris(hydroxyethyl)-s-triazine (MEA triazine) and hexahydro-1,3,5-trimethyl-s-triazine (MMA triazine). In both cases, H_2S reacts with the triazine by means of nucleophilic substitution reactions and a ring-opening, followed by a ring-closing, process. This results in the removal of nitrogen from the ring structure and its replacement with a sulfur atom, forming an amine molecule (**Fig. 52**). The first substitution reaction generates a thiadiazine, and a second substitution reaction is then possible, giving a dithiazine. There is some doubt as to whether a third substitution reaction can occur.

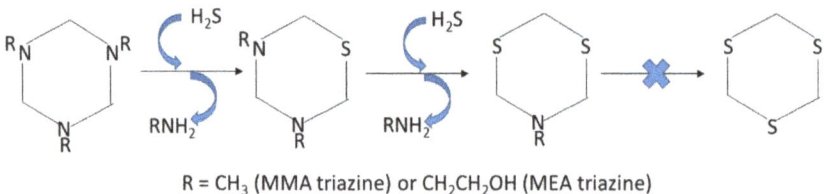

$R = CH_3$ (MMA triazine) or CH_2CH_2OH (MEA triazine)

Fig. 52—Mechanism of triazine reaction with H_2S.

MEA triazine continues to be the most widely used chemical primarily because it is relatively inexpensive and is effective, with good reaction kinetics. It is also more water soluble than MMA triazine, so it tends to be favored for applications when a water phase is present. Triazine products can be prone to the formation of problematic polymeric deposits that tend to accumulate in flowlines and chokes; MMA triazine might offer some benefits over MEA triazine because it is less prone to this problem (Taylor et al. 2017).

Another general problem with triazine scavengers is that their high pH results in a tendency to increase carbonate scaling problems. This problem is often addressed by injecting the scavenger directly into the gas stream so that the scavenger or its reaction products do not directly contact the produced water. An in-line injection quill finely disperses the liquid chemical into the gas stream to maximize the reaction efficiency. However, disposal constraints could mean that it eventually becomes impossible to prevent the mixing of produced water with scavenger reaction products. In such cases, there might be no alternative but to manage the scaling problem through the use of a scale inhibitor (Goodwin et al. 2011), although high inhibitor concentrations might be needed to achieve adequate control.

Calcium carbonate scaling risk is expected to begin at a pH greater than 7.5. This condition might be reached following the addition of approximately 1,000 ppm of sulfide scavenger (Williams et al. 2010). The scaling tendency is also sensitive to small increases in pH. Nevertheless, the actual pH values will be buffered significantly, by both carbonate scale deposition and CO_2 gas dissolution where gas is present.

In addition to exacerbating calcium carbonate scaling risks, H_2S scavengers sometimes induce scaling problems not previously encountered, such as those of zinc or lead sulfide. Therefore, in systems using scavenger injection, it is important to closely monitor the scavenger injection dosage to avoid any overdosing that would worsen the scaling problems. In extreme cases, it could be necessary to use chemical chelating agents to control the problem.

5.7 Cyclic Injection. Cyclic water injection is a process usually used in mature floods where there is a high degree of heterogeneity that results in poor sweep efficiency. The technique has been quite widely deployed in naturally fractured carbonate reservoirs, for example. The concept is extremely simple: It uses alternating, rather than concurrent, periods of injection. For some circumstances, it might be appropriate to consider alternating production as well as injection, although this would entail appreciable production deferment. The technique modifies the existing flood streamlines and thereby has the potential to improve overall flood efficiency. It might have a secondary benefit because water production will be expected to significantly reduce, so there could be a reduced cost associated with the handling of excessive water volumes.

The efficiency of the process is expected to be high in preferentially water-wet rocks. Capillary pressure and relative permeability effects can be responsible for improved cyclic oil displacement at the microscopic level. At the macroscopic flood level, improved sweep of the lower-permeability layers that are in communication with more permeable thief zones might be expected by changing waterflood streamline patterns and alternating the dominance between gravity and viscous forces.

Cyclic injection has been successfully applied in a number of sandstone and carbonate oil fields in Russia and China and in naturally fractured reservoirs in the rest of the world. The process was first mentioned at the beginning of the 1960s.

In Russia in 1961–1962, the Jablonev Ovrag Field was returned to production after having been shut in, and an improvement in oil production rates was observed. Cyclic injection has been widely used in Russian oil fields after positive early results in the Pokrovskoye and Kalinovskoye fields, both carbonate reservoirs (Surguchev et al. 2003).

This technique has been applied widely in the Samara region, Tatarstan, in the Volga-Ural Basin, and western Siberia, with more than 50 fields involved. In Tatarstan, more than 2 000 000 m³ of additional oil was produced in carbonate reservoirs with low- to medium-viscosity oil (up to 64 cp). In total, the cumulative additional oil production by cyclic water injection in these three main oil-producing regions amounted to 23.2 million metric tons by 1984.

In the US, a shock freeze in January 1961 resulted in a 4-day shutdown of injection at the carbonate Spraberry Field in Texas. As a result, oil production increased from 720 to 1,170 B/D, so injection was again shut in a month later. This led to a chain of events with increased oil production and decreasing water production in the months that followed (Elkins and Skov 1963).

Similar positive effects were also observed in the Grayburg–San Andres Formation at the McCamey Field in Texas. When injection rates were reduced from 2,500 to 440 B/D, oil production rates rose from 84 to 270 B/D during the depressurizing phase of cyclic injection. An additional example of a cyclic-injection effect was observed at the fractured Skaggs Pool Field in New Mexico when fluctuations in the oil rate and the injection rate were analyzed during the period 1964–66. Oil rates tended to decline during periods of rising injection rates and to show peaks after injection rates were reduced. These trends led to the conclusion that pressure built up in the fracture system during periods when increased injection rates increased imbibition into the reservoir matrix. The buildup of pressure tended to interfere with the countercurrent flow of oil from the fine pores of the matrix blocks into the fractures where fluid flow was viscous-dominated. However, during periods of reduced injection, pressure in the fracture system declined. With imbibed water then being retained in the matrix by capillary forces, oil could then flow out of the matrix and into the fracture network, where it could be produced.

A key issue when considering cyclic injection is to determine the optimal cycle period. This question was addressed for the Russian Urnenskoe Field (Rublev et al. 2012), where streamline simulation suggested that a cycle time of 5 days of injection followed by a 10-day stop was appropriate. Indeed, most cyclic-injection schemes appear to adopt relatively short cycle times. However, another study based on simulation outcomes (Shchipanov et al. 2008) suggests that incremental recoveries of approximately 3% might be achievable with short-duration cycle times and that benefits of up to 5% might be possible with much longer cycle periods.

Shchipanov et al. (2008), based on the simulation of a Norwegian reservoir, found that incremental benefits associated with cyclic injection and ongoing production were lower than those for alternating periods of production and injection (**Fig. 53**). This makes sense conceptually. After a flood short circuit has been established, relative permeability effects tend to further promote that pathway. Consider, however, a fracture short circuit: Under normal injection conditions, this is pressurized. The pressure sink of the producer tends to promote production from that higher

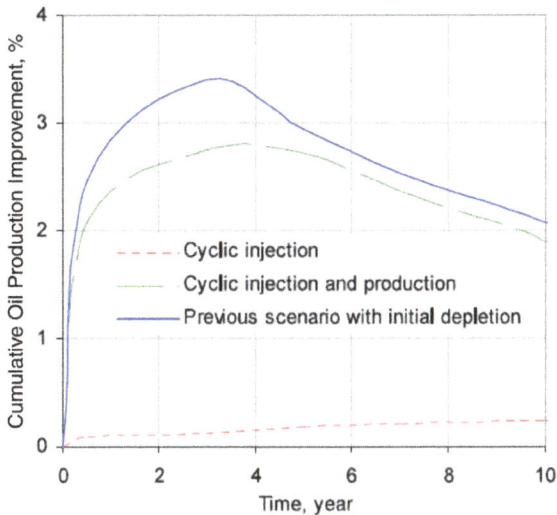

Fig. 53—Cyclic waterflooding scenarios (Shchipanov et al. 2008).

pressure, a tendency that is exacerbated by relative permeability. However, if production is shut in during injection, that pressure sink is removed and the injection will soon redistribute to the matrix. The only downside to this concept is the associated production deferment, so the extra oil that comes from improved sweep must be able to more than offset the production deferment from the injection half-cycle.

The appropriate cycle time introduces an aligned approach in which the injection water is pulsed, although this clearly entails a shorter-term process (minutes rather than days) compared to a standard cyclic-injection scheme (Groenenboom et al. 2003). It has been claimed that this technique can stimulate the reservoir on a large scale through the use of low-frequency pressure waves. Although stimulation of the well appears to have been achieved in Groenenboom et al. (2003), it is not clear there was any overall reservoir stimulation or improvement to the overall flood conformance involved with this technique.

In core experiments designed to assess the benefits and recovery impacts of cyclic injection in the Ekofisk Field, in the Norwegian North Sea (Surguchev et al. 2008), the results indicated that above the bubblepoint, an additional 2.9% of oil originally in place (OOIP) could be produced by cyclic injection. They also showed that an additional 5.9% of OOIP could be generated if the operating pressure was maintained below the bubblepoint but above the critical gas saturation.

The available information therefore suggests this technique is valid to improve recovery in fields with unfavorable heterogeneity in layered sandstone reservoirs, or for carbonate systems where short circuiting resulting from natural fracturing is present.

Cyclic injection could afford benefits even in the absence of capillary, gravity, or heterogeneity effects (Perez et al. 2014). An application of cyclic injection in Las Mesetas Field in Argentina was conducted where the injection distribution between wells and within wells (using injection mandrels) maximized the use of the volumes available from the injection plant, giving a stable total injection volume. This was found to deliver an incremental 11 000 m³ of oil in the first 18 months (**Fig. 54**).

Measured oil production for the project area

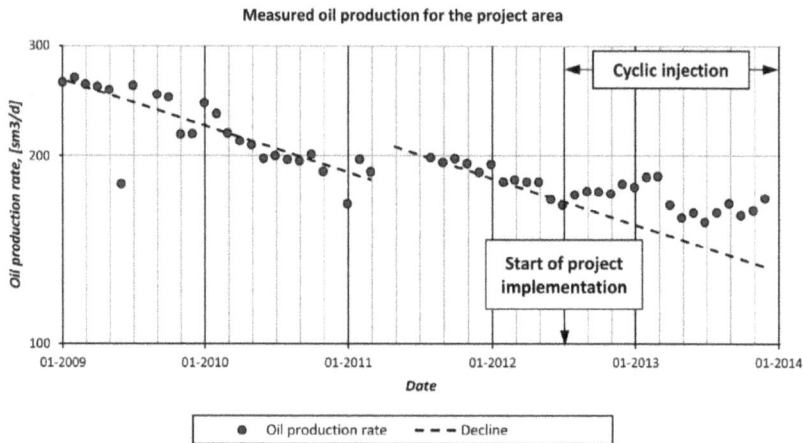

Fig. 54—Oil performance with and without cyclic injection in the pilot area (after Perez et al. 2014).

In summary, cyclic injection appears to be an option with the potential to improve waterflooding for some fields with virtually no additional cost because the only change required is the opening and closing of wells. The total volume of water injected in the reservoir will most likely be reduced under cyclic flooding. However, for many mature fields, the benefits could outweigh the drawbacks associated with lower injection volumes. Cyclic injection appears to have the greatest potential in stratified or highly heterogeneous sandstone and fractured carbonate reservoirs, where improved sweep, accelerated oil production, and reduced water production are the primary positive effects. In very mature fields, there could also be benefits by alternating periods of both production and injection. The improved sweep in such cases must be able to more than offset the periods of production deferment.

The cyclic-injection process induces cycling of pressure. This could create unstable conditions in zones of different oil saturations. It has been suggested that the existing capillary nonequilibrium on the contacts of oil-saturated and water-saturated zones is weakened (Surguchev et al. 1992). In a water-wet reservoir, the imbibition of water in a zone of high oil saturation is accelerated during the first half-cycle by a positive pressure gradient. Then, during the second half-cycle when the pressure decreases, the imbibed water displaces a volume of oil into the more permeable zone and imbibition is accelerated as a result. Consequently, the additional oil is released and displaced from the previously nonswept regions.

Separate injection in different wells in different phases or in the same well at different times are techniques that might offer sweep benefits by accelerating crossflow between zones of low and high oil saturations.

5.8 Further Development. A range of potential additional development options are available for fields that have been developed under waterflood. The options, and the circumstances in which they might be useful, are discussed in the sections that follow.

5.8.1 Infill. Infill developments can range from the drilling of individual wells to infilling complete patterns to a smaller well spacing. In both cases, it is likely that the oil delivered entails elements of both acceleration and incremental recovery. It is very surprising that the recovery benefits from infill were not widely recognized until a combination of infill and the commencement of waterflooding was proposed in the late 1970s in the US for large carbonate reservoirs in New Mexico and west Texas to overcome reservoir heterogeneity and improve flood continuity (van Everdingen and Kriss 1980).

It was subsequently reported that substantial reserves were added by means of infill in nine fields (seven of which were waterfloods) in Texas, Oklahoma, and Illinois and that an incremental recovery of 10 000 m^3 of oil was achieved for each infill well (Barber et al. 1983). These applications demonstrated that infill could deliver economically viable volumes in waterfloods. A review of the success of infill in Texas waterfloods subsequently confirmed that the benefits were derived from the improvement in well connectivity it afforded (Reviere and Wu 1986). This suggests that the infill option might be equally applicable at any stage of field development.

For the infill of pattern developments, it is possible to infill across the entire field. However, when the STOIIP density is assessed, it might sometimes be the case that infill will only be justified at the crest or in specific parts of the field. It has been suggested that infill drilling might be particularly applicable to waterfloods in heavy-oil systems because the required pressure drop can be too high over long distances, leading to the concept of short-distance oil displacement (Turta and Singhal 2000).

When considering infill, it is important to assess the well spacing that can be justified on the basis of the incremental oil volumes that might be delivered and the associated cost, given the number of new wells required, because eventually a point will be reached where the incremental volumes are no longer justified by the incremental costs. It is also important to assess whether it is possible to achieve an infill pattern on any particular spacing from what has already been installed in the field. For example, standard infilling of a seven-spot pattern will normally result in subsequent well spacings that are one-third of the original well spacings. Such projects are rarely justified. Indeed, the literature does not seem to include any such examples in a waterflood, although there is an example in a steamflood development (Belghache et al. 2016). **Fig. 55** illustrates the number of additional wells that would be required in such an endeavor.

It is also possible to perform infill with horizontal waterflood patterns. The infill of a horizontal pattern development on a 120-m well spacing down to a 60-m spacing development was planned for a carbonate reservoir in Oman **(Fig. 56)** (Garimella et al. 2012). This reduction in well spacing was expected to increase recovery from the initial 25% up to 45%. The differential in recovery between the different well spacings was understood at the time of the initial development, but it was a conscious decision to begin development on the larger well spacing to appraise field performance and gather information in support of the long-lead surface facilities needed to handle the increased volumes associated with the full-field infill development.

Fig. 57 shows the variation in ultimate recovery expected for patterns in different parts of this field. This illustrates the importance of well spacing on recovery.

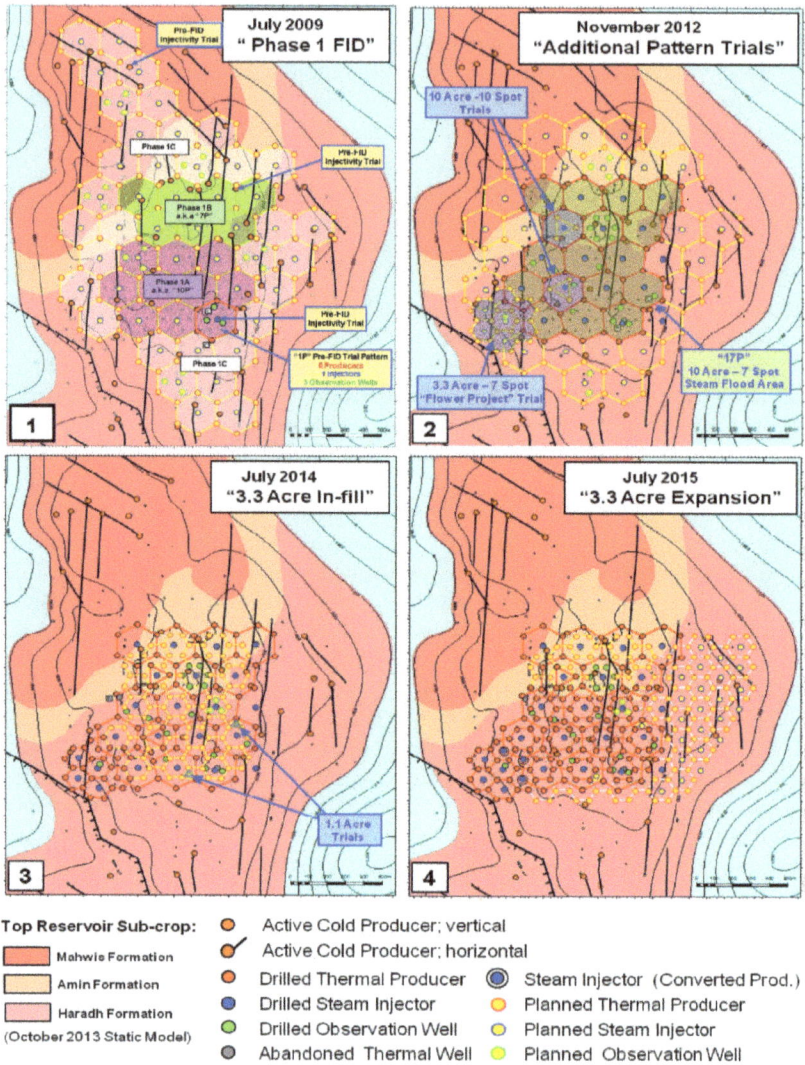

Fig. 55—Infill of a seven-spot pattern (Belghache et al. 2016). FID = final investment decision.

However, the rock type and presumably the thickness of the oil column also have important effects on recovery. In some fields, these factors can influence whether infill is justified in different parts of the field.

In some low-permeability reservoirs, infill might sometimes be needed just to obtain injector-to-producer communication within an adequate time frame. In other words, infill drilling is sometimes needed to get the waterflood to work at all. Alternatively, infill in low permeabilities might help improve recovery by removing the negative effects of geology. In the Ordos Basin in China, where average permeabilities

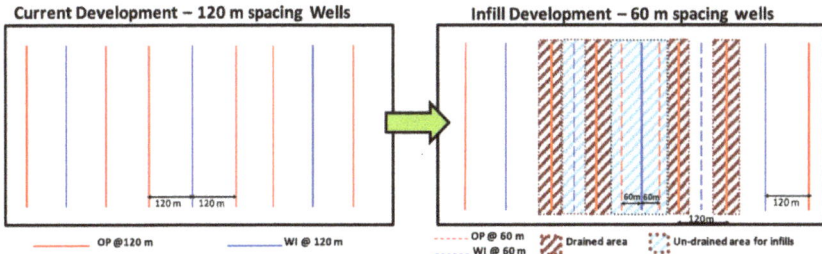

Fig. 56—Infill in a horizontal waterflood development (Garimella et al. 2012). OP = oil producer; WI = water injector.

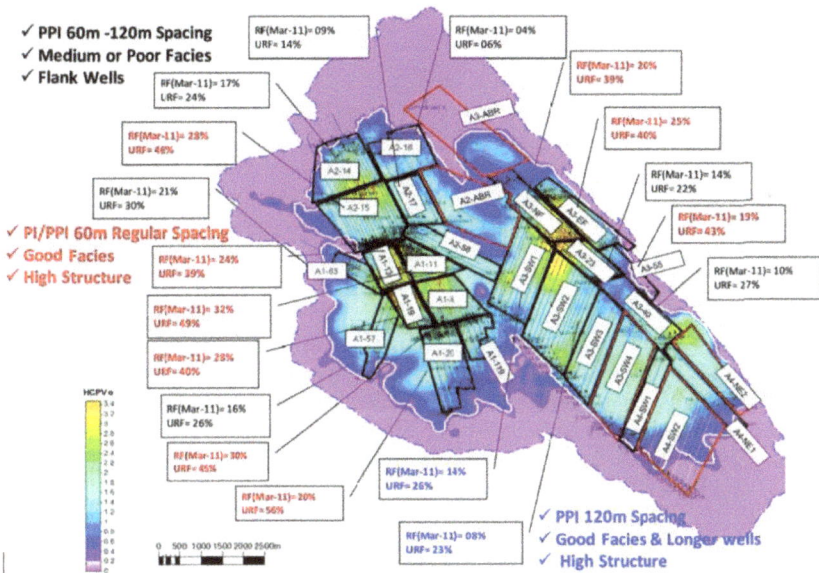

Fig. 57—Impact of well spacing on recovery factor (RF) (Garimella et al. 2012). PI = producer-injector; PPI = producer-producer-injector; URF = ultimate recovery factor; HCPV = hydrocarbon pore volume.

range from 0.3 to 10 md, characterization of the geological factors responsible for flood short circuiting was the key to the optimization of subsequent well infill (Yu et al. 2018). Well data were analyzed to quantify which wells performed adequately under waterflood, which received no waterflood support, and which saw support but suffered premature water breakthrough. This latter category was clearly the primary factor negatively impacting flood performance. More detailed analysis showed that dynamic fracture propagation was responsible for the lack of sweep. The primary direction of water breakthrough was in line with the direction of maximum horizontal in-situ stress: 70° northeast.

The field was developed using a nine-spot pattern. Five scenarios were considered (**Fig. 58**). The analysis showed that an infill development in which the flood would

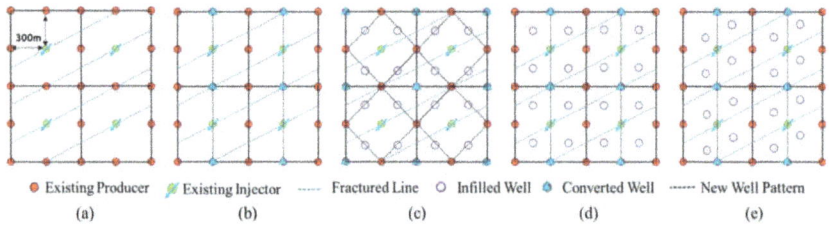

Fig. 58—Infill drilling options for a tight reservoir in the Ordos Basin (Yu et al. 2018).

be converted to a staggered line drive aligned with the in-situ stress orientation to limit the impacts of induced fracture growth was the best option (Fig. 58e). After this change was made, the average oil production rates increased from 1.6 to 1.8 t/d, producing water cut dropped from 55 to 49%, and extrapolation of performance data suggested an increase in recovery from 20 to 25%.

When infill is justified on an individual-well basis, there is sometimes a tendency to drill producers at the expense of injectors. Therefore, it is important that injection and voidage requirements are always considered. Because infill entails the drilling of extra wells, there should also be an assessment of whether the production and injection volumes can be handled through the existing processes. In the event that such capacity is not available, the project might still be viable, but the costs of providing that additional capacity will need to be borne by the project, which could make the economics more challenging.

An additional challenge with infill for onshore pattern developments results from the significant numbers of new wells and the associated flowlines introducing significant complexity because the infrastructure suddenly becomes much more closely spaced. Rig movement and well accessibility can be particularly problematic areas. This issue was recently addressed for a very congested field development in Oman (Hussain et al. 2019). An initial assessment suggested that only 30% of the planned infill wells were drillable considering the infrastructure already in place. Consequently, an integrated multidisciplinary team with expertise in concept engineering, geomatics, production geology, reservoir engineering, and well engineering was formed, to assess the urban planning issues. A fresh approach to the way wells were drilled meant that it was subsequently possible to drill 90% of the locations compared to the 30% deemed possible in the original assessment. This enabled the maturation of the planned target hydrocarbon volume and created significant value in the redevelopment of this field. It was estimated that appropriate urban planning might save approximately 10% in off-plot capital expenditure. Furthermore, urban planning helped in lowering HSE risks during drilling and also reduced production deferment during construction.

One of the aims of an infill project is to improve the recovery factor for a field because a waterflood has the potential to leave appreciable volumes of oil in the reservoir. An alternative to an infill is to apply a tertiary recovery process. The appropriate scheme would depend on how much oil remains in place, which option would recover the largest proportion of the remaining volumes, and the relative cost of the competing projects. It is therefore appropriate to consider these alternative options when considering further development. However, the more efficiently the initial waterflood performs, the lower the target remaining for these competing projects will be and thus the less likely it will be that any incremental project will be economically justified. The best of the enhanced oil recovery (EOR) options will depend on the reservoir properties. These options will be discussed in the sections that follow.

Alusta et al. (2011) described a methodology for the assessment of infill against chemical EOR. Alusta et al. (2012) discussed the economic factors that might impact the subsequent outcome, but these might be expected to be different for different cases. Of course, one of the potential scenarios could be that both infill and EOR are viable for subsequent field optimization. Indeed, this is the outcome arising from a study on a deepwater turbidite (Armih et al. 2013).

5.8.2 Chemical Enhanced Oil Recovery. One of the options for further waterflood development is to convert the water source such that a low-salinity flood is implemented (see *Waterflooding: Chemistry*, another book in this series). However, low-salinity flooding is best used early in field life for maximum effect, so it might not be competitive for application as a late-life EOR process. This section will therefore focus on chemical EOR options, which tend to be the type of EOR used when the oil is somewhat viscous, meaning the displacement efficiency of a normal waterflood is not particularly good.

Historically, there have been two main types of chemical EOR:

- Polymer injection, which is a process that works by increasing water viscosity and improving the oil/water mobility ratio
- Surfactant injection, which changes the oil/water interfacial tension (IFT), leading to significantly lower residual oil saturations after the flood

There is, however, an additional level of granularity in chemical EOR options because, in addition to the polymer option, there are options for surfactant/polymer flooding to reduce the oil/water IFT, a surfactant flood to modify wettability, and combinations of these such as alkaline/surfactant/polymer (ASP) flooding.

All of these options entail the addition of rather expensive chemicals to the injection water. They are therefore typically applied as tertiary recovery processes following a period of waterflood that is used to better understand the displacement process so the expensive EOR process is applied efficiently.

Polymer. Polymer flooding involves the injection of water in which the viscosity has been enhanced by the addition of a polymer. Polymer flooding is therefore one of the few EOR techniques that does not reduce the amount of trapped oil resulting from capillary forces. Instead, it accelerates the time required for the flood to reach the residual oil saturation because of the improvement to the mobility ratio. Increasing the water viscosity also suppresses the viscous fingering that might otherwise occur when waterflooding with high-viscosity oils and improves the water/oil shock front. Consequently, there will be improved displacement at reduced water cuts and a lower swept-zone oil saturation for a given volume of water injected. So, in polymer flooding, the remaining oil saturation decreases, but the residual oil saturation does not.

Although polymer flooding improves sweep along the waterflood displacement pathways, it has limited potential to develop areas of bypassed oil. This is because the relative permeability of well-swept water zones is higher than the relative permeability of unswept zones at initial oil saturation. This option therefore faces the same challenges of conformance control that are faced by a conventional waterflood.

There has been some debate regarding whether a polymer might improve displacement for heterogeneous reservoirs. However, the polymer will typically be just as negatively impacted by heterogeneity as normal water. Therefore, polymer flooding generally starts to become economically attractive as the oil viscosity increases. It is usually the least expensive EOR technique to apply, but the incremental recoveries also tend to be the lowest of the EOR techniques, typically ranging from 2 to 10%.

Polyacrylamides and polysaccharides are the two main types of polymers used in polymer flooding. The performance of polyacrylamides is a function of the molecular weight and the degree of hydrolysis. This latter parameter defines the ionic character of the polymer and is controlled by replacing some of the acrylamides with acrylic acid or converting some of them to acrylic acid (depending on the manufacturing process) before polymerization (**Fig. 59**).

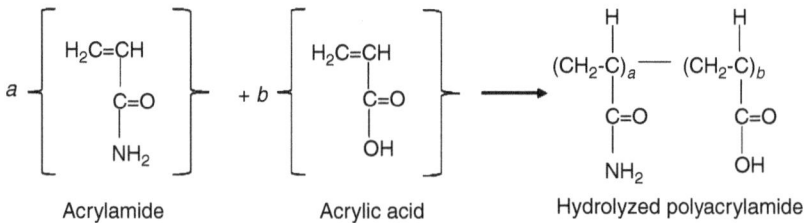

$$a \left[\begin{array}{c} H_2C=CH \\ | \\ C=O \\ | \\ NH_2 \end{array} \right] + b \left[\begin{array}{c} H_2C=CH \\ | \\ C=O \\ | \\ OH \end{array} \right] \longrightarrow \begin{array}{cc} H & H \\ | & | \\ (CH_2\text{-}C)_a & (CH_2\text{-}C)_b \\ | & | \\ C=O & C=O \\ | & | \\ NH_2 & OH \end{array}$$

Acrylamide   Acrylic acid   Hydrolyzed polyacrylamide

Fig. 59—Formation of hydrolyzed polyacrylamide.

The main factors affecting polyacrylamide performance are the salt content of the brine, reservoir temperature, and adsorption. The degree of hydrolysis in the reaction in Fig. 59 is given by $b/(a+b)$. The polyacrylamides used in polymer-injection projects usually have a degree of hydrolysis in the range of 25–30%. Some forced hydrolysis can occur when the polymer is injected into the reservoirs, so the injected product might have a degree of hydrolysis closer to 25% than 30%. This range is selected to avoid the dangers of polymer precipitation from a reaction with divalent cations. Increased levels of hydrolysis increase the number of negatively charged carboxyl groups in the polymer chain. This consequently increases the degree of repulsion between the polymer molecules and negatively charged rock, reducing the polymer adsorption.

Polymer adsorption has two main effects. First, it reduces the permeability of the rock, which is beneficial to flood performance because it makes the mobility ratio between the slug and chasing water more favorable. However, it also delays the breakthrough of the polymer slug.

The degree of adsorption depends on several factors such as the type of polymer and solvent conditions. The higher the adsorption on the rock, the more polymer that needs to be injected, increasing the cost of the flood. There is an extra factor—hydrodynamic acceleration—that is a result of the velocity profile in the pore. The large polymer molecules cannot approach the walls of the pore closely, and they therefore tend to move in the higher-velocity region at the center of the pore. This causes the polymer molecules to travel, on average, faster than the average speed of the water phase, resulting in earlier polymer breakthrough. This results in a displacement efficiency reduction because hydrodynamic acceleration results in a spreading of the polymer over a larger volume. Consequently, the polymer concentration decreases and the mobility of the polymer fluid increases.

The polymer will have a tendency to ball up in the presence of electrically conductive saline water solutions, which might require the injection water to have a low salinity. Polysaccharides are not as susceptible to this effect as polyacrylamides. Polyacrylamide molecules are charged along their backbones and, because the charges repel each other, the molecules take the shape of a large, fluffy, ball. The larger the volume of the molecules, the higher the viscosity of the water for a given concentration. Ions in the water tend to neutralize the charges on the polymer molecules, and therefore, the ball contracts, reducing the viscosity. Thus, to obtain the same viscosity, the amount of dissolved polymer must be increased.

High temperatures cause polymers to rapidly hydrolyze and precipitate, and the temperature at which this occurs will be a function of the divalent-ion concentration. Therefore, testing is needed to establish the exact levels of stability to be expected in any application. In general, however, problems tend to begin to be observed at temperatures greater than 100°C.

The mechanism for polysaccharides (which are sometimes called biopolymers) is different than that for polyacrylamides. They form long, stiff rods and thereby increase the water viscosity. Although charges are involved in the molecules, the creation of a double or triple helix in the molecular structure makes the molecules very stiff and almost insensitive to salinity. However, polysaccharides are readily attacked by many types of bacteria, and this attack can break up the chains and so reduce, or even destroy, the viscosifying power. An even greater drawback is their cost, which can be up to an order of magnitude higher than that of polyacrylamides.

Most historical polymer projects were pilots to evaluate the technology, but in recent years, there have been a number of full-field deployments. These include Daqing in China (Dong et al. 2008), the world's largest polymer flood. This polymer project has boosted the expected recovery to more than 50%, which is 10–12% more than would have been achieved using waterflood alone. Favorable conditions for polymer flooding at Daqing include the in-situ oil viscosity, ranging from 6 to 9 cp, and a formation-water salinity ranging from 3000 to 7000 mg/L. The findings from the extensive experience in this field include

- It is important to understand the degree of heterogeneity to assess the viability of a polymer flood. A permeability ratio of less than 5 was deemed to be necessary within each unit for the polymer to be effective.
- Higher-molecular-weight polymers, a broad range of polymer molecular weights, and higher polymer concentrations are beneficial in the injected slugs.
- Surveillance is important to optimization, including the monitoring of the polymer mixing process.
- At the start of a polymer project, the producing water cut will initially continue to rise until the effects of the polymer are seen. This will then decrease and then stabilize for a period before later water-cut increases inevitably occur.

Polymer flooding has also been successful in Oman's Marmul Field, where the in-situ oil viscosity is 90 cp (**Fig. 60**) (Al-Saadi et al. 2012). The initial deployment has been more successful than originally anticipated, and consequently, a phased deployment is being implemented.

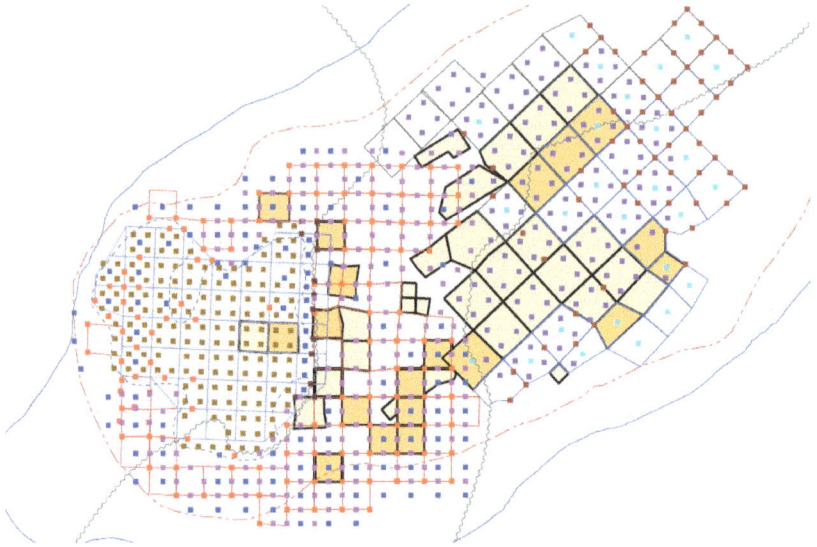

Fig. 60—Polymer deployment architecture at Marmul (Al-Saadi et al. 2012).

In all deployments, it seems that piloting has been a critical precursor to demonstrate the viability of this development philosophy. At Marmul, the polymer injectivity was one of the key uncertainties. In practice, this turned out to be much higher than originally anticipated. The full-field deployment entails the use of a combination of five- and nine-spot patterns. The five-spot patterns use a 400-m spacing, with later plans to infill to a nine-spot (Huseby et al. 2016).

Water quality must be very closely controlled in a polymer flood to ensure polymer viscosity is properly maintained. The water quality specifications set for the first phase of a polymer development at the Mangala Field in India are (Pandey et al. 2012)

- Iron: < 5 ppm
- Total suspended solids: < 2 ppm
- Oil content: Nil
- Dissolved oxygen: < 10 ppb

Such specifications require a very high degree of operational focus and control.

Surfactant/Polymer. Surfactant/polymer tends to be an alternative to ASP injection (see the next section) that can be considered when an inexpensive freshwater source is not available (the alkaline part of ASP requires fresh water to avoid scale precipitation) or where ASP cannot be used because of consumption reactions with gypsum or anhydrite within the reservoir.

The surfactant/polymer process tends to be economically challenging compared to that of ASP because the alkaline in an ASP process reduces surfactant adsorption, thereby reducing the use of an expensive surfactant.

The surfactant can be designed to either reduce water/oil IFT or to modify reservoir wettability. The most common surfactants used for EOR are sulfates and sulfonates, which are anionic surfactant derivatives of sulfuric acid. When these surfactant solutions contact an oil phase, the surfactant has a tendency to accumulate at the interface. As a result, only small surfactant concentrations are needed to saturate that interface.

The objective of an IFT-reducing surfactant flood is to mobilize a portion of the oil that has been left behind after the displacement achieved by a waterflood. In a conventional waterflood, the point will eventually be reached in the parts of the reservoir contacted by injected water where capillary forces trap the remaining oil, rendering it immobile. Part of this oil might be mobilized by reducing the IFT between the two phases. Typically, IFT needs to be reduced by three to four orders of magnitude (from initial values of approximately 10) to achieve any material effect. IFT reduction causes the flow between the phases to become more miscible, allowing the phases to move more easily in the presence of each other.

The molecules of the surfactants used consist of a hydrophilic head (for example, sulfonate or carboxylate) and a lipophilic tail (typically a hydrocarbon chain). The incorporation of both these features within a single molecule means they will arrange themselves at the interface between the oil and water phases. The hydrophilic heads will be in the water phase, and the lipophilic tails will be in the oil phase. This encourages the formation of micelles: aggregations of one phase in another (**Fig. 61**). It is possible to have oil micelles in water, or vice versa.

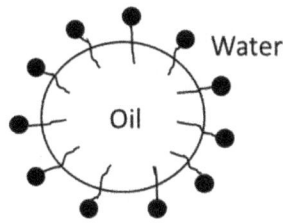

Fig. 61—Oil micelle in water.

The surfactant concentration at which micelle formation commences is known as the critical micelle concentration (CMC), and the surfactant concentration must be greater than this CMC to have appropriate properties for mobilizing oil. The CMC of surfactant increases significantly at higher temperatures. This therefore becomes an important design parameter that is likely to impair project economics in unfavorable temperature conditions (Noll 1991).

One of the key effects of surfactant molecules is the creation of a force that competes with the effect of IFT. The surfactant makes the phases more miscible and enables the water and oil to move more easily in each other's presence. This means that the relative permeability curves will be clearly influenced. A straightening of the curves reflects the more miscible flow between the phases. In addition to a recovery benefit associated with the changes in the shapes of the relative permeability curves, there might also be benefits associated with a reduction in the residual oil saturation. (In **Fig. 62**, the original curves are solid lines and the surfactant-modified curves are dashed.)

A second way in which surfactants can influence flood recovery is through the modification of rock wettability. This effect is likely to be particularly important in cases where the reservoir is relatively oil-wet, and surfactants are then applied to make the rock more water-wet. Making such changes has two key effects: The water endpoint relative permeability is materially decreased, and the residual oil saturation is increased (**Fig. 63**). Because this process reduces the theoretical ultimate recovery benefits, it could require early application so the positive impact on the fractional flow curve from lowering the water endpoint relative permeability can be realized. However, the reduction in ultimate recovery will have a limited impact because of the effects of discounting.

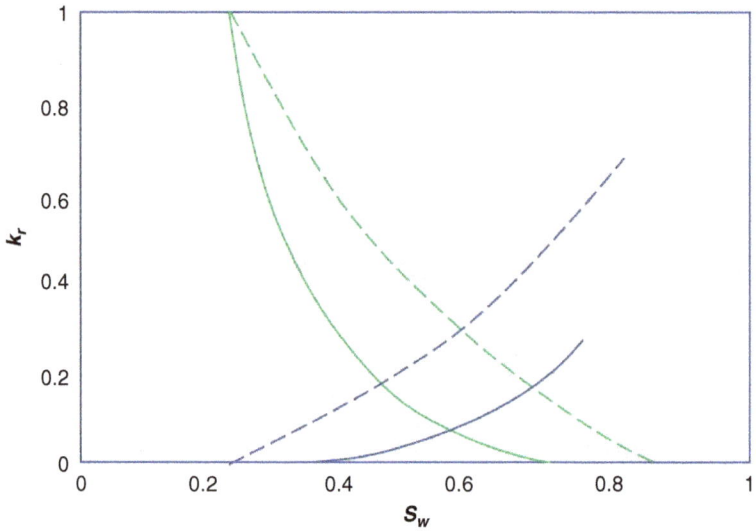

Fig. 62—Impact of surfactant on relative permeability curves.

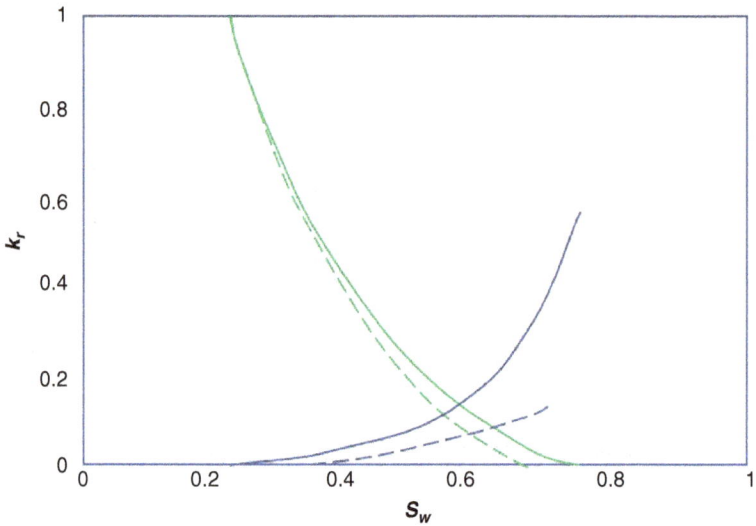

Fig. 63—Impact of wettability modification on relative permeability curves.

In surfactant flooding, a relatively small slug (generally less than 0.3 pore volumes) of concentrated surfactant solution (5–10% by weight of active matter) is typically applied. This is followed by a larger slug of polymer solution (generally greater than 0.6 pore volumes) designed for proper mobility control. The surfactant slug affords ultralow oil/water IFT, which is needed to mobilize some of the residual oil from waterflooding. The low IFTs needed to achieve this are usually accompanied by the formation of a microemulsion, a third phase that is in equilibrium with the oil and water phases at the same time. These microemulsions can be relatively viscous, even with low-viscosity crude oils. This results in a mobility requirement for the polymer buffer that is significantly higher than that required for the surfactant slug alone.

A blend of at least two surfactant types is usually needed. The primary surfactant is either a petroleum sulfonate or a synthetic aromatic sulfonate. Secondary surfactants are selected from a large range of available anionic and nonionic types such as alcohol ethoxylates and alcohol ethoxy sulfates. Cosolvents, usually low-molecular-weight alcohols, are also added to reduce the formation of gels, liquid crystals, and macroemulsions.

One disadvantage of using alcohol is that it decreases the solubilization of oil and water in microemulsions, therefore increasing the minimum value of IFT achievable with a given surfactant. For low-temperature reservoirs, the need for the cosolvent effect of alcohol can be reduced, or even eliminated, by using surfactants with branched hydrocarbon chains, adding ethylene oxide and/or propylene oxide to the surfactant, or using a blend of two dissimilar surfactants, because the single-phase region is larger for a surfactant blend than for either surfactant alone.

Petroleum sulfonates possess a number of features that led them to be considered for use as the surfactant in this type of flood, not least of which have been their relatively low cost and their availability in large quantities. The structure of the surfactant is very important, and the degree of branching within the molecule is known to play a particularly important role (with some degree of branching being beneficial).

It is the ionic sulfonate group in the petroleum sulfonate molecule that confers the ability to reduce IFT. However, it is this very characteristic that tends to induce precipitation in high-salinity-formation-brine environments. This problem has led to the development of a range of other surfactants such as sulfated or sulfonated ethoxylated fatty alcohols or alkyl phenols, which display both anionic and nonionic characters within the same molecular structure, and alpha olefin sulfonates.

Alkyl benzene sulfonates have also been promoted as attractive surfactants. These surfactants, synthesized with a narrow molecular-weight range, have been applied in the Daqing Field (Wu et al. 2008). More details on the range of surfactants available and their structures can be found in Levitt et al. (2009).

Alkaline/Surfactant/Polymer. Chemical flooding using alkali and surfactants injected together with polymers improves the waterflood displacement through the reduction of the capillary forces that trap oil so that the residual oil saturation is reduced. The trapped oil gives rise to a residual oil saturation that limits the ultimate recovery achievable in waterflooding projects. Many methods have been used in attempts to lower the IFT between oil and water, reducing the residual oil saturation after flooding.

The first attempts at lowering capillary forces involved the injection of surfactants to lower the IFT, as described in the previous section. However, although this showed promise in laboratory experiments, early field trials were disappointing because only small pore volumes of surfactant could be economically injected in the field because of their high costs. This problem steered the evolution of surfactant flooding toward alkaline flooding.

This process lowers the surface tension between the oil and water interface by creating petroleum soaps or surfactants in situ through the reaction of inexpensive alkali (which must be added to the injected fluid) with the acid components in the crude oil (a process called saponification). One benefit of in-situ surfactant creation is that there will be a reduction in the adsorption of anionic surfactants on the reservoir rocks (Nelson et al. 1984). The challenge in this process is to optimize the chemical behavior of the system, taking into account the salinity of the formation brine while maintaining the optimal chemistry. Alkaline consumption by the rock significantly influences the flow of a dilute alkaline solution through the reservoir. This creates the dilemma of needing to choose between the highest displacement efficiency, which is achieved at the lowest IFT, and an acceptable displacement rate. The need to make that decision is removed by the addition of a cosurfactant that raises the concentration of electrolytes needed for minimum IFT to alkali concentrations high enough for adequate propagation of the alkaline bank. In other words, the cosurfactant raises the salinity needed for an alkaline flood.

In early work with this technology, the most commonly used alkali was sodium hydroxide (NaOH). A wide range of alternatives—including sodium orthosilicate (Na_4SiO_4), sodium carbonate (Na_2CO_3), sodium metasilicate (Na_2SiO_3), ammonium hydroxide (NH_4OH), ammonium carbonate [$(NH_4)_2CO_3$], and sodium metaborate ($NaBO_2$)—has also been used. The alkali can induce materially significant scaling and corrosion problems. It can also have a detrimental effect on polymer performance, so additional polymer might be needed to achieve the required viscosity so the selected alkali might therefore need to find a balance between all of these effects.

Sodium hydroxide has the advantage that only small amounts are needed to influence the pH. However, it can induce severe silicate scaling problems through reactions with reservoir rock. It also performs poorly compared to sodium carbonate in that it induces a much greater degree of surfactant adsorption. Unfortunately, sodium carbonate is a much weaker alkali, so much larger volumes of chemical are needed if this product is selected.

A comparison between these two alkalis has been made in the Daqing ASP development. Guo et al. (2016) found that while the recovery achievable with both alkalis was similar, with sodium carbonate the producing water-cut reduction was more significant, the produced-fluid processing was easier, and the scaling problems were fewer. Consequently, it seems that at Daqing, there was a preference for using sodium carbonate as the alkali.

Alkali also readily reacts with divalent cations, which will further induce scaling problems. For example,

$$Na_2CO_3 + Ca^{2+} \rightarrow 2Na^+ + CaCO_3. \quad \dotfill \quad (4)$$

Apart from the problems of the scale itself, the consumption of the alkali in such reactions is a significant problem. Therefore, water softening is usually incorporated in the process, reducing the divalent-ion concentrations to less than 10 ppm.

A number of different components in the crude oil can be involved in reactions with the injected alkali, including carboxylic acids and, to a lesser extent, carboxyphenols, porphyrins, and asphaltenes. A generic form for the reaction of alkali with acidic components in the crude is given as

$$NaOH \text{ (alkali)} + RCOOH \text{ (acid crude)} \rightarrow RCOONa \text{ (surfactant)} + H_2O \text{ (water)}. \quad (5)$$

Although injected alkali has a beneficial impact on reducing IFT, there is an associated negative impact because it tends to reduce solution viscosity. Because the alkali also undergoes a number of reactions with minerals dissolved in the formation water, and with the reservoir rocks themselves, the optimal alkali concentration becomes somewhat of a balancing act. For this reason, a buffer such as sodium carbonate is applied. In Daqing, the only full-scale project so far, the alkali concentration injected ranged from 0.6 to 1.4 wt% (Guo et al. 2016). Average ASP parameters for low-salinity sandstone reservoirs can be in the following range:

- Alkali: Average wt% = 1.0%.
- Surfactant: Average wt% = 0.3% (although sometimes more could be needed).
- Polymer: Average wt% = 0.1% (although this depends on the oil viscosity, so requirements should be more accurately determined on the basis of the viscosity of the emulsions formed in the mixing of the ASP solution with the oil).
- Average ASP flood volume = 0.35 pore volumes.

For heterogeneous reservoirs, it has been suggested that the viscoelasticity of the solution could be more important than IFT reduction (Hou et al. 2001a, 2001b). This implies that improvements to large-scale displacement could be more important than pore-scale effects. If that is the case, there could be opportunities to reduce the concentration of alkali. This would widen the range of surfactants that could be used as well as reduce the appreciable scaling problems experienced in ASP floods. Field testing of this concept in Daqing is described by Wan et al. (2006).

The injected surfactant and in-situ generated surfactant are chemically rather different. However, they share the general property that their interfacial activity depends on the environment, in particular the salinity. Indeed, there is an optimal brine salinity at which the surfactants reduce the oil/water IFT most strongly. This phenomenon is related to the phase behavior of the surfactant or the soap (Hou et al. 2001a).

In the under-optimum regime, both surfactants preferentially partition into the brine phase. In the over-optimum regime, they partition into the oil phase, forming viscous oleic emulsions. This condition tends to prevent the oil from moving farther through the reservoir toward the producers. It is only near their respective optimal salinities that the surfactants are able to generate a third, separate microemulsion phase, which exhibits very low IFTs with both the water and the oil phases. The optimal salinity is surfactant-specific. The optimal salinity of the surfactants generated by the alkali reactions with the crude is generally significantly lower than that of typical injected surfactants. The optimal salinity of the system as a whole will be a combination of the optimal salinities of both surfactants.

This implies that, for chemical flooding, it is preferable to design the flood such that the optimal salinity of the chemical slug is at or close to the actual brine salinity to achieve a low oil/water IFT. However, because alkali is being added in the flood, this also confers salinity, so the optimal salinity will be somewhat higher. Failure to meet this condition will cause the surfactants to be either flushed ineffectively to the producing wells (under-optimum case) or partitioned into immobile oil—that is, to be retained and thus lost (over-optimum case). The formation of the in-situ surfactant at the front of the ASP slug will create an activity gradient that has a function similar to that of the salinity gradient.

Unfortunately, because of dispersion and mixing, controlling the chemical concentrations might not be feasible. One possible remedy applicable to the case of surfactant/polymer flooding is the establishment of a salinity gradient. If the initial brine salinity in the reservoir is higher than the optimal salinity of the surfactant, a dilution of the surfactant resulting from dispersion is reduced because any surfactant molecule traveling ahead of the front of the chemical slug is retarded as it enters an over-optimum regime.

At the rear end of the chemical slug, the higher viscosity reduces the dispersion. These two effects thus confine the surfactant slug and limit dilution. At low initial reservoir salinities, on the other hand, this concept is not applicable. An alternative, though more expensive, remedy would be to increase the volume of the chemical slug.

The ASP technique can remedy the inherent problems of both surfactant/polymer flooding and alkaline flooding. Alkali, surfactant, and polymer are injected together as one slug. In this slug, the alkali and the surfactant will generally travel at slightly different velocities. The surfactant is subject to partitioning into any remaining oil and to matrix adsorption. The alkali is consumed by the saponification, the precipitation of carbonates, and possibly exchange reactions with the matrix. The ASP design must therefore be robust such that the slug does not disintegrate as a result of chromatographic separation.

As injected, the phase behavior of the ASP slug will be under-optimum. Then, upon contact of the alkali with crude oil, the saponification process leads to a reduction of the optimal salinity in the region where soap is generated such that the phase behavior locally changes to over-optimum. As a consequence, a gradient in optimal salinity from over-optimum at the front of the chemical slug to under-optimum at the rear of the chemical slug is established. This serves to confine the chemical slug and limit dispersive dilution. This inherent gradient can progress through the reservoir, moving, in principle, an optimal interfacially active zone through the reservoir that leaves no oil behind.

A second benefit of the alkali is its effect on surfactant adsorption. By turning the surface charge of the rock more negative at higher pH values, the adsorption of anionic (negatively charged) injected surfactants is greatly reduced. A drawback of the injection of alkali is the increase of a scaling potential at the producing wells if the reservoir contains substantial amounts of divalent ions such as calcium and magnesium.

The effect of the polymer in stabilizing the displacement is still very important. Unless oil liberated by a chemical slug is displaced by means of an internal polymer drive (or, in the case of steeply dipping reservoirs, potentially gravity segregation), this oil will be retrapped as soon as the interfacially active region has passed. One optional component typically added to an ASP mixture is a solvent that promotes the common solubilization of both surfactant and polymer in the brine phase. Additional benefits of the solvent are a faster phase equilibration of the ASP mixture when it is brought in contact with the oil and a reduction of the viscosity of the microemulsion phase. The solvents are usually low- to medium-molecular-weight alcohols.

One of the important requirements for ASP flooding is good compatibility between surfactant and polymer. As previously stated, the ASP solution contains the water-soluble surfactant, and therefore there is excellent compatibility between the polymer and the surfactant. The mixture should be clear and stable. Diluted surfactant solutions have little effect on the viscosity of polyacrylamide polymers, but at high surfactant concentrations, or low temperatures, changes in phase are often observed.

When the brine contains a polymer, a condensed phase (or precipitation) occurs at a low surfactant concentration because of the exclusion of the polymer from the microemulsion phases. However, at a low surfactant concentration, the microemulsion phases are small, and there should be enough water for the polymer to dissolve. Large viscosities are detrimental to oil recovery because they can cause local viscous instabilities during a displacement or decreased injectivity.

The shock-front mobility ratio of an ASP flood should be near 1, making the end-point mobility ratio range between 3 and 7. In practice, the flood can be preceded by a preflush of fresher water to mitigate the unfavorable effect of ions in the formation water. This is followed by the injection of the ASP slug to mobilize oil, and then the slug is driven through the reservoir with a graded polymer drive.

A large number of tests were conducted in the US that helped to mature this complex technology (Mayer et al. 1983). A successful field trial was carried out in the White Castle Field in Louisiana (Falls et al. 1994). Two enhanced alkaline slugs, a carbonate slug and a silicate slug, were developed for the White Castle pilot. In the carbonate slug, sodium carbonate was the only alkali. In the silicate slug, sodium carbonate still provided most of the alkalinity, but some sodium silicate was also included in the formulation. The following provides a brief description of the slug components and their functions:

- Sodium carbonate: Provides alkalinity for converting petroleum acids to soaps. It puts a negative charge on the rock, which reduces surfactant adsorption and so minimizes wasteful reactions with reservoir brine and rock.
- Sodium silicate: In addition to providing the same functions as sodium carbonate, it can improve the filtration characteristics of injectants, reduce gravel-pack dissolution, and protect surfactants from interaction with magnesium in the reservoir.
- Cosurfactant: Raises the salinity requirement of the system to a level where alkali concentrations are high enough to travel through the reservoir without appreciable delay.
- Sodium chloride: Adjusts activity of the system relative to the region of optimal salinity. Can fine tune the system by changing the concentration.
- Surfactant: Keeps the slug single phase at the ambient temperature without significantly affecting performance at the reservoir temperature.
- Soft water: Hardness ions, such as calcium and magnesium, precipitate under alkaline conditions, so the makeup water has to be softened.

The volume of oil produced in the White Castle pilot surpassed expectations. However, the test identified problems associated with carbonate scaling in the producers and tight emulsions associated with the produced fluids.

To avoid problems with ASP floods, close attention must be paid to proper slug formulation. Important aspects to consider include

- The compatibility of the injected slug with formation water should be checked to minimize issues associated with productivity impairment and tight emulsion formation.
- A large number of chemicals are used in an ASP process, so particular attention should be paid to ensure the compatibility of all the chemicals needed and

to verify their compatibility with the polymer. (Polymer degradation can occur as a result of overdosing.) The chemicals normally required will include an oxygen scavenger, a biocide, a corrosion/scale inhibitor, and pH control additives.

- The biocide must be added to the makeup water before polymer mixing because bacterial growth in the polymer mixing units could otherwise occur. This is particularly important in projects using a biopolymer.

Field experience suggests that surface-related problems can be difficult to manage in ASP floods. Problems experienced include

- Incomplete polymer mixing resulting in viscosity loss.
- Injectivity issues during polymer-slug injection. This problem has been seen in a number of applications.
- Polymer shear degradation effects resulting in additional polymer concentration requirements.
- Corrosion problems observed in production systems.
- Oil/water separation difficulties have been reported for most projects. The extent of the problems has been such that even demulsifiers have struggled to resolve the problems.
- Significant scaling problems, of which alkali is the source. When sodium hydroxide is used as the alkali, these scaling problems include silicate scales that are very difficult to treat.

ASP work has historically focused on applications to sandstone reservoirs with low salinity. Subsequently, researchers developed surfactants that expanded applications to higher formation-water salinities (Barnes et al. 2008) and carbonate reservoirs (Adibhatla and Mohanty 2006).

Polymer vs. Alkaline/Surfactant/Polymer Flooding. When an ASP flood follows a conventional waterflood, the incremental recoveries can be approximately 10–25% of STOIIP. An ASP field test at Daqing recovered an additional 20% of the oil-in-place from that recovered by waterflood (Shutang and Qiang 2010). By comparison, a polymer flood might normally deliver an incremental recovery in the range of 2–10% of STOIIP. However, although the recovery benefits of an ASP flood are clearly better, the process is more complex, and it is by no means certain that the economics for the ASP option will be better.

In fields where the in-situ viscosity is low (< 10 cp), the expected incremental recovery of a regular polymer flood over that achieved in a standard waterflood is small. This could make the ASP option the preferred one for such circumstances. Conversely, when the incremental recovery afforded by a polymer flood is appreciable (i.e., when the in-situ viscosity is higher), the lower costs of polymer flooding could make this a more attractive proposition than ASP.

5.8.3 Water-Alternating-Gas Injection. Water-alternating-gas (WAG) injection is another option to increase oil recovery above what can be achieved by waterflooding alone. It is a process that combines some of the attractive features of both water injection and gas injection. The gas/oil system has a lower IFT than the oil/water system. Therefore, the microscopic displacement of the oil by gas is normally better than by water, so gas injection tends to result in a much lower residual oil saturation. Unfortunately, although it gives better microscopic displacement, the high gas mobility, because of its low viscosity, causes gas fingering and early gas break-through, which reduces the macroscopic sweep. Alternating water injection with gas injection is an effective method of reducing gas permeability. This enables an improvement in the microscopic sweep as a result of the gas injection while limiting the decrease in macroscopic sweep because of high gas mobility. This leads to an improvement in the overall recovery (**Fig. 64**).

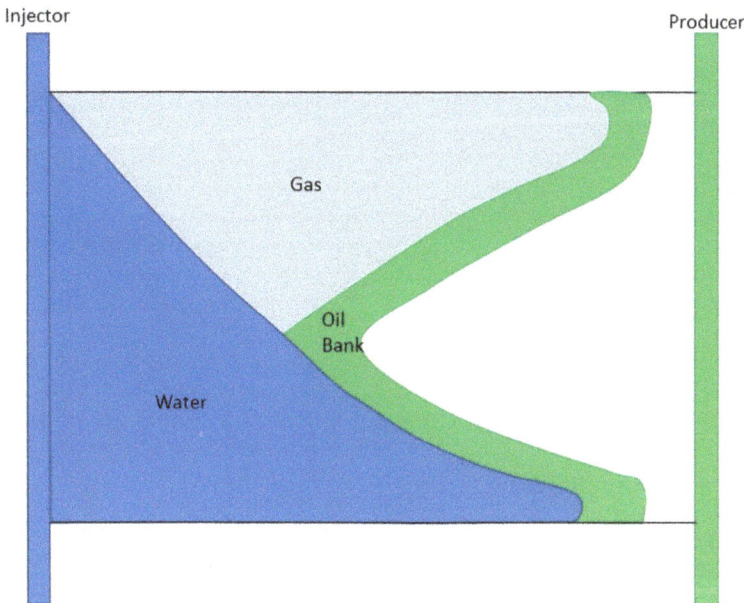

Fig. 64—WAG scheme.

There are a number of different types of WAG.

In miscible WAG (mWAG), the injection pressure in the gas cycle is greater than the minimum miscibility pressure of the reservoir fluid. The miscible front in the mWAG process has poor volumetric sweep efficiency because of its low viscosity. However, when the injected gas is miscible with the oil, the gas can completely dis-solve in the oil and displace virtually all the oil in the reservoir, achieving a very high recovery efficiency. Without miscibility, the gas tends to bypass the oil, leaving much of it behind, so there is an appreciable value associated with achieving miscibility. Sometimes there can be a drive to repressurize, initially using water, to achieve those required miscibility conditions before deploying the WAG process.

At lower pressures, the gas will be immiscible with the oil. The primary purpose of performing immiscible WAG (iWAG) is to improve frontal stability and contact unswept zones. The displacement efficiency of an iWAG development is thus much lower than that expected for mWAG (Ma and Youngren 1994), although the residual oil saturations in the gas-flooded zones will be reduced. The main benefits are likely to come from opportunities to improve the macro sweep efficiency resulting from the architecture of the flooded reservoir, such as fining up reservoir sequences and attic fault traps. Other benefits of iWAG include significant voidage replacement, delayed waterflood breakthrough, and well hydraulic benefits of in-situ gas lift.

In a simultaneous WAG (SWAG) process, the water and gas are injected together. This eliminates the need for separate water- and gas-injection lines, therefore reducing operational costs. In cases where the export of gas is not economical, reinjection of the produced gas in a SWAG scheme can significantly reduce, or in certain cases eliminate, the need for flaring.

In some fields, WAG has been implemented by means of a large slug of gas injection followed by a small number of slugs of water and gas. This is known as hybrid WAG injection. In a tapered WAG project, the length of the injected gas cycles is reduced after the initial gas-injection cycle. This provides more incremental oil for a given injection volume of gas. It might therefore be appropriate to sequentially perform tapered WAG in different parts of the field.

To optimize the WAG efficiency and improve recovery, the amount of gas and water injected into the oil reservoir must be optimized. Large amounts of gas injection result in front instability and a poor macroscopic sweep efficiency, while too much water injection will reduce microscopic efficiency. There tends to be a reduction in incremental recovery as the number of cycles increases, which limits the number of cycles used in WAG projects.

The first WAG project was conducted in the North Pembina Field in Canada in 1957 (van Poolen 1980). A review of projects conducted up until 2001 (Christensen et al. 2001) showed that WAG is nearly always deployed as a tertiary recovery scheme. The review also showed that mWAG projects tend to predominate over iWAG projects. Although WAG projects have been used in a wide range of permeabilities, there is a predominance of projects in higher permeabilities.

The sizes of the initial gas slug are mostly in the range of 0.1–3 pore volumes. The initial WAG ratio used (defined as the volume of water injected in a cycle divided by the volume of gas injected in the cycle) is typically 1, although this can range up to 3 or more. There is an approximate 50% split between projects that use hydrocarbon gas and projects that use CO_2 as the gas source, with only a small number of projects using other gas sources.

There are a number of useful features associated with CO_2 that result in its extensive use in WAG projects. First, because it is a greenhouse gas, its use in recovery processes helps limit environmental impacts. Second, at reservoir conditions with reservoir pressures greater than 1,200 psi and temperatures less than 250°F, CO_2 can be miscible with a wide range of crude-oil types. By comparison, a hydrocarbon solvent enriched with the C_2–C_4 components can have a miscibility pressure in the range of 1,500–3,000 psi for mid-API-gravity oils.

Although WAG projects offer clear recovery benefits over standard waterflooding, which has led to wide deployment of this technology, there are a number of potential problems that might need to be managed when using this process. These include

- Early gas breakthrough: Several field cases have reported early gas breakthrough resulting from channeling or override. In offshore projects, gas injection can be an attractive way to manage the produced gas, but if that gas cycles through the reservoir quickly, it could result in the field's gas compression capacity becoming constrained. Such problems are difficult to solve and could result in wells being shut in far earlier than originally anticipated.

- Pressure loss: In miscible projects, loss of pressure can be a very serious problem because loss of miscibility will result in a significantly lower recovery.
- Reduced injectivity: Changes in relative permeability because of three-phase flow can result in reduced injectivity, as can hydrate- or asphaltene-deposition problems. This, in turn, has the potential to result in reduced reservoir pressures, which might cause problems related to miscibility. Reduced injectivity of water is often observed following the injection of a gas slug. The precipitation of asphaltenes is likely to be a problem near the wellbore, but relative permeability effects are more likely to be far-field effects. These problems can occur for a number of reasons including water relative permeability reduction and the trapping and bypassing of gas. It might be possible to manage such problems by decreasing the WAG ratio, thereby decreasing mobility control. Hadlow (1992) suggests that, on average, a 20% loss in water injectivity will be seen after CO_2 injection.
- Asphaltenes (Khanifar et al. 2011): Several fields have experienced problems as a result of asphaltenes because gas can have the effect of decreasing the concentration of resins that help to stabilize the asphaltenes present in the crude oil. There are several instances of problems in field operations of CO_2 miscible floods. It might be possible to treat such problems using solvent treatments.
- Hydrate formation: Hydrates form when gas molecules become trapped inside a cage of water molecules. They can form at high pressures and low temperatures (typically below 20–25°C). During a gas-injection cycle at Ekofisk, an injection well became completely plugged within a matter of hours (Hermansen et al. 1997).
- Corrosion: This is a very common problem in WAG projects. It is often seen in injection systems where existing facilities are used for a tertiary WAG development. Problems are more likely to occur when CO_2 is used as the gas-injection source (Bonis 2012).
- Scale: Scale problems are most likely to occur when CO_2 is the gas-injection source.

Whether a WAG process is economically viable sometimes depends on whether the available gas has any inherent value. Under most situations, development costs (per barrel of oil recovered) of a WAG process are higher than those of waterflooding. However, there are special situations when WAG injection could be attractive:

- The hydrocarbon gas is stranded because of the lack of a local gas market.
- The gas contains high concentrations of CO_2, which make it a lower-value resource.
- A pure, or nearly pure, CO_2 source is available. This factor has been the driver for the application of many CO_2 WAG projects in the US.

At lower pressures, depending on the degree of solvent enrichment, the gas will be immiscible with the oil. This means that the injected gas forms a separate phase from the oil phase. It therefore tends to bypass the oil because of its much higher mobility and transits the reservoir quickly, where it is produced in the producing wells. This means that the displacement efficiency of an iWAG scheme (Christensen et al. 2001) is much lower than that of the mWAG scheme. However, there will still be benefits associated with a residual gas saturation reduction in those parts of the reservoir that have been contacted by the injected gas.

5.9 Blowdown. As a waterflood matures, it is inevitable that the water cut will increase, resulting in a progressive reduction in the net oil production. Eventually, it will no longer be economical to continue to spend money injecting water to support the declining produced oil volumes. As this time approaches, there could be an additional development option to consider for mature waterfloods.

The waterflood focus is to maximize the value and recovery of oil present in the reservoir. In so doing, gas production is curtailed because the reservoir pressure is maintained. For locations with an available gas market, a point will eventually be reached when the economics of the remaining oil that can be produced can no longer compete with the value of the gas. At that time, water injection can be stopped to allow the reservoir pressure to drop, thereby liberating solution gas for production. The volume of gas that might be liberated in such a blowdown process will need to be assessed to confirm that this represents an economic proposition. Gas might be liberated from both unswept and swept parts of the reservoir. It then migrates to the crest of the structure, where it is produced (**Fig. 65**).

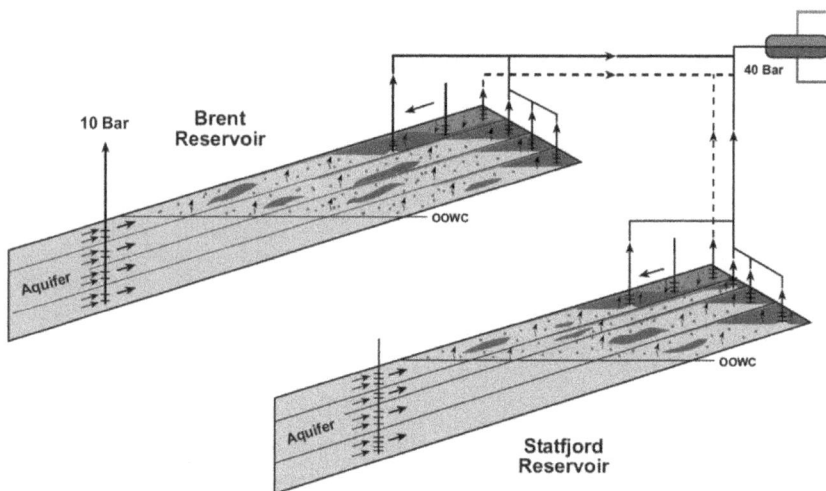

Fig. 65—Reservoir blowdown process (Braithwaite 1994). OOWC = original oil/water contact.

Depressurization was undertaken for the Brent Field, in the North Sea. This process was expected to release approximately 1.5 Tcf of additional gas (Braithwaite 1994). Furthermore, the studies also predicted the process would increase oil and condensate recovery by 34 million bbl compared to a continuation of the waterflood.

Before the project could be initiated, uncertainties such as aquifer influx, reservoir souring, compaction, subsidence, and sand production had to be assessed and contingency measures identified. The gas-cap growth in the process depends on the speed of the pressure decline. In the Brent process, any excess gas liberated that was not needed for the gas sales contract was to be stored in the top layers of both Brent and Statfjord, where the gas cap would be expanded to a predetermined volume and additional oil recovery could be realized through a gravity drainage process (Fig. 65).

Drainage of both oil and condensate helps sustain the oil rims, and careful planning of midflank well locations is required to maximize oil recovery.

The blowdown process is controlled by the back production of water from the reservoir. The criteria to be managed include gas production and availability, oil production, water production, gas injection, water injection, pressure, and gas-cap size in each reservoir unit and area of the field (Gallagher et al. 1999).

The overall efficiency of the process is governed by the gas saturation at which the liberated solution gas becomes sufficiently mobile to flow to the production wells (critical gas saturation) and by the ultimate gas saturation at the end of the depressurization. It is evident that a detailed understanding of critical gas saturation and relative permeability is vital to understanding just how efficient the process will be (Egermann and Vizika 2000). Studies related to these factors have been reported for Brent depressurization (Ligthelm et al. 1997). The importance of a proper understanding of gas behavior under the process is therefore critical (Goodfield and Goodyear 2003). Studies indicate that depressurization experiments can result in residual gas saturations as high as 40% unless the experiments are run slowly (as would occur in the real process). Goodfield and Goodyear (2003) showed that capillary and gravity forces are likely to dominate the displacement. It is also important to incorporate considerations of compaction to avoid the overprediction of blowdown benefits (Tehrani et al. 2002).

Schemes similar to that applied in Brent were also investigated for Miller (Beecroft et al. 1999) and South Brae (Drummond et al. 2001) in the North Sea. At the Statfjord Field, the process was expected to result in an increase in oil recovery from 65 to 68% and an increase in gas recovery from 53 to 74%. These measures were expected to extend field life by 10 years (Boge et al. 2005).

6. Conclusions

Waterfloods rarely, if ever, perform as originally expected. This invariably occurs because the knowledge about the field, and the geology in particular, is incomplete at the time that development is sanctioned. To achieve an optimal field performance, therefore, it is inevitable that some degree of remediation will be needed at some time. The remediations are likely to include activities that improve the displacement efficiency delivered by the wells and the completions that were initially installed. However, a number of supplementary activities could be needed to secure longer-term production. Activities in this category include the remediation of corrosion problems, modifications of facilities to handle increased water cut, and treatments to reduce produced H_2S concentrations, for example.

Remediation activities are unlikely to be optimized unless robust knowledge is available regarding the exact nature and location of the problem. This requirement demands that all waterflood projects incorporate a fit-for-purpose surveillance program. That program will need to be tailored to the key uncertainties associated with any given development. Furthermore, as the understanding of the project changes so will the information that is needed. As a consequence, therefore, the surveillance program will typically need to be updated on an annual basis.

In many cases, the remediation activities will lead to an overall field performance that is better than that at the time development was originally sanctioned, although in other cases, the improvements will only bring performance back up to that originally predicted. In these latter cases, it is often the case that the original predictions were unrealistic, usually because the reservoir complexity was underestimated. In many cases, full-field optimization will entail further development through the addition of wells. This can range from the drilling of ad-hoc additional wells to drain additional volumes to the drilling of entire infill patterns. In some cases, the additional development might entail some change to the type of recovery process, and there are many examples of waterfloods being converted into chemical enhanced oil recovery or water-alternating-gas processes. Again, robust data are needed to provide a sound foundation such that an optimized decision is implemented.

Waterfloods are complex, and it is unlikely that the initial development will lead to an optimized field performance. Robust surveillance programs thus play a key role in understanding the many moving parts such that longer-term performance can be optimized.

7. Nomenclature

ABC	=	after before compare
ASP	=	alkaline/surfactant/polymer
B	=	relative permeability ratio parameter
bbl	=	barrel
B_o	=	oil formation volume factor
BOD	=	biological oxygen demand
BOPD	=	barrels of oil per day
B_w	=	water formation volume factor
BWPD	=	barrels of water per day
CDG	=	colloidal dispersion gel
CFD	=	computational fluid dynamics
CMC	=	critical micelle concentration
CO	=	carbon oxygen
CO_2	=	carbon dioxide
COD	=	chemical oxygen demand
cp	=	centipoise
CRM	=	capacitance resistance modeling
DGGE	=	denaturant gradient gel electrophoresis
D_{HI}	=	Hall plot derivative
EM	=	electromagnetic
EOR	=	enhanced oil recovery
ER	=	electrical resistance
ESP	=	electrical submersible pump
E_v	=	volumetric sweep efficiency
EZIP	=	expandable zonal inflow profiler
FBA	=	fluorinated benzoic acid
FISH	=	fluorescent in-situ hybridization
GAB	=	general aerobic bacteria
GAnB	=	general anaerobic bacteria
GOR	=	gas/oil ratio
HC	=	hydrocarbon
HCPV	=	hydrocarbon pore volume
H_2S	=	hydrogen sulfide
HSE	=	health, safety, and environment

ICD	=	inflow control device
ICV	=	inflow control valve
IFT	=	interfacial tension
I_H	=	Hall integral
in.	=	inch
IR	=	infrared
iWAG	=	immiscible water alternating gas
km	=	kilometer
KPI	=	key performance indictor
k_r	=	relative permeability, md
k_{ro}	=	relative permeability to oil, md
L	=	liter
LPR	=	linear polarization resistance
m	=	meter
mah	=	meters along hole
MAR	=	microautoradiography
MBOPD	=	thousands of barrels of oil per day
md	=	millidarcy
MEA triazine	=	hexahydro-1,3,5-tris(hydroxyethyl)-s-triazine
mL	=	milliliter
MMA triazine	=	hexahydro-1,3,5-trimethyl-s-triazine
MPN	=	most probable number
mWAG	=	miscible water alternating gas
n	=	molar ratio
nm	=	nanometer
OBC	=	ocean-bottom cable
OIIP	=	oil initially in place
OOIP	=	oil originally in place
OOWC	=	original oil/water contact
OP	=	oil producer
PCA	=	principal component analysis
PCR	=	polymerase chain reaction
P_e	=	reservoir pressure, psi
PI	=	producer-injector
PLT	=	production logging tool
PNL	=	pulsed neutron log
ppb	=	parts per billion
PPI	=	producer-producer-injector
ppm	=	parts per million
psi	=	pounds per square inch
PV	=	pore volume
P_{wf}	=	flowing bottomhole injection pressure, psi
qPCR	=	quantitative polymerase chain reaction
RF	=	recovery factor
rRNA	=	ribosomal RNA
SNL	=	spectral noise logging
SRB	=	sulfate-reducing bacteria
STOIIP	=	stock-tank oil initially in place

S_w	=	water saturation, %
SWAG	=	simultaneous water alternating gas
SWB	=	seawater breakthrough
SWCTT	=	single-well chemical tracer test
t	=	time
t_D	=	ratio of cumulative liquid production (Q_L) to the total pore volume of the pattern area
TDT	=	thermal decay time
THPS	=	tetrakis(hydroxymethyl)phosphonium sulfate
UAE	=	United Arab Emirates
UK	=	United Kingdom
URF	=	ultimate recovery factor
US	=	United States
USA	=	United States of America
VDPM	=	viscous disproportionate permeability modifier
VRR	=	voidage replacement ratio
WAG	=	water alternating gas
WHP	=	wellhead pressure
W_i	=	volume of water injected, m^3
WI	=	water injector
WOR	=	water/oil ratio
WOR'	=	water/oil ratio time derivative
Y	=	oil fraction function
μm	=	micron

8. References

Abbad, M., Dyer, S., Ligneul, P. et al. 2015. In-Line Water Separation: A New Promising Concept for Water Debottlenecking Close to the Wellhead. Paper presented at the SPE Middle East Oil & Gas Show and Conference, Manama, Bahrain, 8–11 March. SPE-172687-MS. https://doi.org/10.2118/172687-MS.

Abdulhadi, M., Kueh, P. T., Zamanuri, A. et al. 2018. Cost Effective Water Shut-Off: Slickline Conveyed High Expansion Ratio Through Tubing Bridge Plug. Paper presented at the SPE Asia Pacific Oil and Gas Conference and Exhibition, Brisbane, Australia, 23–25 October. SPE-192145-MS. https://doi.org/10.2118/192145-MS.

Addoun, M., Maraf, K., Tighe, M. et al. 2011. Initial Deployment of a New Design One-Trip Straddle System Resolves High Pressure Cross-Flow Problem in Water Injection Well in Algeria. Paper presented at the SPE/ICoTA Coiled Tubing & Well Intervention Conference and Exhibition, The Woodlands, Texas, USA, 5–6 April. SPE-143336-MS. https://doi.org/10.2118/143336-MS.

Adibhatla, B. and Mohanty, K. K. 2006. Oil Recovery from Fractured Carbonates by Surfactant-Aided Gravity Drainage: Laboratory Experiments and Mechanistic Simulations. Paper presented at the SPE/DOE Symposium on Improved Oil Recovery, Tulsa, Oklahoma, USA, 22–26 April. SPE-99773-MS. https://doi.org/10.2118/99773-MS.

A/Fotuh, M. and Macary, S. 2000. Factors That Affect the Success of Mechanical Water Shut-off in Wells. Paper presented at the SPE Annual Technical Conference and Exhibition, Dallas, Texas, USA, 1–4 October. SPE-62891-MS. https://doi.org/10.2118/62891-MS.

Ahmed, Q. A., Ibrahim, A., Salah, R. et al. 2010. Risk Analysis and Decision Making in Relative Permeability Modifier Water Shut-off Treatment. Paper presented at the North Africa Technical Conference and Exhibition, Cairo, Egypt, 14–17 February. SPE-126845-MS. https://doi.org/10.2118/126845-MS.

Al-Bimani, A. S., Al-Sharji, H. H., Aihevba, C. O. et al. 2006. Enhancing Oil Production from Mature Fields by Focusing on Well-Intervention Management: North Oman. Paper presented at the SPE/ICoTA Coiled Tubing Conference & Exhibition, The Woodlands, Texas, USA, 4–5 April. SPE-99706-MS. https://doi.org/10.2118/99706-MS.

Allan, M. E., Rahman, M., and Reed, D. A. 2013. The Challenges of Full Field Implementation of Fiber-Optic DTS for Monitoring Injection Profile in Belridge Field, California. Paper presented at the SPE Digital Energy Conference, The Woodlands, Texas, USA, 5–7 March. SPE-163694-MS. https://doi.org/10.2118/163694-MS.

Al-Mahrooqi, M. A., Marketz, F., and Hinai, G. 2007. Improved Well and Reservoir Management in Horizontal Wells Using Swelling Elastomers. Paper presented at the SPE Annual Technical Conference and Exhibition, Anaheim, California, USA, 11–14 November. SPE-107882-MS. https://doi.org/10.2118/107882-MS.

Al Mahrooqi, S. and Azim, M. K. 2015. Solving a Mysterious Casings Failure Problem in ESP Fitted Production Field. Paper presented at the SPE Middle East Oil & Gas Show and Conference, Manama, Bahrain, 8–11 March. SPE-172736-MS. https://doi.org/10.2118/172736-MS.

Al-Saadi, F. S., Al-Amri, B. A., Al Nofli, S. M. et al. 2012. Polymer Flooding in a Large Field in South Oman—Initial Results and Future Plans. Paper presented at the SPE EOR Conference at Oil and Gas West Asia, Muscat, Oman, 16–18 April. SPE-154665-MS. https://doi.org/10.2118/154665-MS.

Al Saidi, A., Pourafshary, P., and Al Wadhahi, M. 2015. Application of Fast Reservoir Simulation Methods to Optimize Production by Reallocation of Water Injection Rates in an Omani Field. Paper presented at the SPE Middle East Oil & Gas Show and Conference, Manama, Bahrain, 8–11 March. SPE-172633-MS. https://doi.org/10.2118/172633-MS.

Al-Shahrani, F. F., Baluch, Z. A., Al-Otaibi, N. M. et al. 2007. Successful Water Shut-off in Open Hole Horizontal Well Using Inflatables. Paper presented at the SPE Saudi Arabia Section Technical Symposium, Dhahran, Saudi Arabia, 7–8 May. SPE-110968-MS. https://doi.org/10.2118/110968-MS.

Alusta, G. A., Mackay, E. J., Fennema, J. et al. 2011. EOR vs. Infill Well Drilling: How to Make the Choice? Paper presented at the North Africa Technical Conference and Exhibition, Cairo, Egypt, 20–22 February. SPE-150454-MS. https://doi.org/10.2118/143300-MS.

Alusta, G. A., Mackay, E. J., Fennema, J. et al. 2012. EOR vs. Infill Well Drilling: Sensitivity to Operational and Economic Parameters. Paper presented at the North Africa Technical Conference and Exhibition, Cairo, Egypt, 20–22 February. SPE-150454-MS. https://doi.org/10.2118/150454-MS.

Amedu, J. and Nwokolo, C. 2013. Improved Well and Reservoir Production Performance in Waterflood Reservoirs – Revolutionizing the Hall Plot. Paper presented at the SPE Nigeria Annual International Conference and Exhibition, Lagos, Nigeria, 5–7 August. SPE-167602-MS. https://doi.org/10.2118/167602-MS.

API RP 45, Recommended Practice for Analysis of Oilfield Waters. 1998. Washington, DC: API.

Armih, K., Alusta, G. A., Mackay, E. J. et al. 2013. Decision Making Tool to Assist in Choosing Between Polymer Flooding and Infill Well Drilling: Case Study. Paper presented at the SPE Enhanced Oil Recovery Conference, Kuala Lumpur, Malaysia, 2–4 July. SPE-165276-MS. https://doi.org/10.2118/165276-MS.

Bangkong, S. A., Alrabaei, R. A., Al-Ajmi, M. D. et al. 2017. Successful Debottlenecking of a Surface Injection Facility. Paper presented at the SPE Kingdom of Saudi Arabia Annual Technical Symposium and Exhibition, Dammam, Saudi Arabia, 24–27 April. SPE-188108-MS. https://doi.org/10.2118/188108-MS.

Barber, A. H., George, C. J., Stiles, L. H. et al. 1983. Infill Drilling to Increase Reserves—Actual Experience in Nine Fields in Texas, Oklahoma, and Illinois. *J Pet Technol* **35** (8): 1530–1538. SPE-11023-PA. https://doi.org/10.2118/11023-PA.

Barnes, J. R., Smit, J., Smit, J. et al. 2008. Development of Surfactants for Chemical Flooding at Difficult Reservoir Conditions. Paper presented at the SPE Symposium on Improved Oil Recovery, Tulsa, Oklahoma, USA, 20–23 April. SPE-113313-MS. https://doi.org/10.2118/113313-MS.

Barnette, J. C., Copoulos, A. E., and Biswas, P. B. 1992. Acquiring Production Logging Data with Pulsed Neutron Logs from Highly Deviated or Non-Conventional Production Wells with Multiphase Flow in Prudhoe Bay, Alaska. Paper presented at the SPE Western Regional Meeting, Bakersfield, California, USA, 30 March–1 April. SPE-24089-MS. https://doi.org/10.2118/24089-MS.

Beecroft, W. J., Mani, V., Wood, A. R. O. et al. 1999. Evaluation of Depressurisation, Miller Field, North Sea. Paper presented at the SPE Annual Technical Conference and Exhibition, Houston, Texas, USA, 3–6 October. SPE-56692-MS. https://doi.org/10.2118/56692-MS.

Belghache, A., Al-Hinai, S., Rabaani, A. et al. 2016. Challenges and Learnings from Thermal Development of Thick Heavy Oil Reservoirs in Southern Oman. Paper presented at the SPE EOR Conference at Oil and Gas West Asia, Muscat, Oman, 21–23 March. SPE-179812-MS. https://doi.org/10.2118/179812-MS.

Bennet, D. and Hoffmann, H. 2018. Oilfield Microbiology: Molecular Microbiology Techniques Used During a Biocide Evaluation. Paper presented at the Offshore Technology Conference Asia, Kuala Lumpur, Malaysia, 20–23 March. https://doi.org/10.4043/28411-MS.

Blount, C. G., Copoulos, A. E., and Myers, G. D. 1991. A Cement Channel-Detection Technique Using the Pulsed-Neutron Log. *SPE Form Eval* **6** (4): 485–492. SPE-20042-PA. https://doi.org/10.2118/20042-PA.

Boge, R., Lien, S. K., Gjesdal, A. et al. 2005. Turning a North Sea Oil Giant into a Gas Field—Depressurization of the Statfjord Field. Paper presented at the SPE Offshore Europe Oil and Gas Exhibition and Conference, Aberdeen, UK, 6–9 September. SPE-96403-MS. https://doi.org/10.2118/96403-MS.

Bonis, M. 2012. Managing the Corrosion Impact of Dense Phase CO2 Injection for an EOR Purpose. Paper presented at the Abu Dhabi International Petroleum Conference and Exhibition, Abu Dhabi, UAE, 11–14 November. SPE-161207-MS. https://doi.org/10.2118/161207-MS.

Bourgoin, S., Sobirin, M., Mahardhini, A. et al. 2014. Clad Through Clad: A Mechanical Water Shut Off Solution for Commingled Production from Multilayered Reservoirs. Paper presented at the International Petroleum Technology Conference, Doha, Qatar, 19–22 January. IPTC-17610-MS. https://doi.org/10.2523/IPTC-17610-MS.

Bourne, H. M., Williams, G., and Hughes, C. T. 2000. Increasing Squeeze Life on Miller with New Inhibitor Chemistry. Paper presented at the International Symposium on Oilfield Scale, Aberdeen, UK, 26–27 January. SPE-60198-MS. https://doi.org/10.2118/60198-MS.

Braithwaite, C. I. M. 1994. A Review of IOR/EOR Opportunities for the Brent Field: Depressurisation, the Way Forward. Paper presented at the SPE/DOI Improved Oil Recovery Symposium, Tulsa, Oklahoma, USA, 17–20 April. SPE-27766-MS. https://doi.org/10.2118/27766-MS.

Bui, T. D. and Jalali, Y. 2004. Diagnosis of Horizontal Injectors with Distributed Temperature Sensors. Paper presented at the SPE Annual Technical Conference and Exhibition, Houston, Texas, USA, 26–29 September. SPE-89924-MS. https://doi.org/10.2118/89924-MS.

Callegaro, C., Masserano, F., Bartosek, M. et al. 2014. Single Well Chemical Tracer Tests to Assess Low Salinity Water and Surfactant EOR Processes in West Africa. Paper presented at the International Petroleum Technology Conference, Kuala Lumpur, Malaysia, 10–12 December. IPTC-17951-MS. https://doi.org/10.2523/IPTC-17951-MS.

Canbolat, S. and Parlaktuna, M. 2012. Well Selection Criteria for Water Shut-Off Polymer Gel Injection in Carbonates. Paper presented at the Abu Dhabi International Petroleum Conference and Exhibition, Abu Dhabi, UAE, 11–14 November. SPE-158059-MS. https://doi.org/10.2118/158059-MS.

Carvajal, G. A., Boisvert, I., and Knabe, S. 2014. A Smart Flow for SmartWells: Reactive and Proactive Modes. Paper presented at the SPE Intelligent Energy Conference & Exhibition, Utrecht, Netherlands, 1–3 April. SPE-167821-MS. https://doi.org/10.2118/167821-MS.

Chan, K. S. 1995. Water Control Diagnostic Plots. Paper presented at the SPE Annual Technical Conference and Exhibition, Dallas, Texas, USA, 22–25 October. SPE-30775-MS. https://doi.org/10.2118/30775-MS.

Cheung, S. K. and Quilter, R. M. 1998. Design and Implementation of a Gel Treatment to Shut Off Water in a High-Temperature, High-Pressure Fractured Reservoir. Paper presented at the SPE Annual Technical Conference and Exhibition, New Orleans, Louisiana, 27–30 September. SPE-49075-MS. https://doi.org/10.2118/49075-MS.

Christensen, J. R., Stenby, E. H., and Skauge, A. 2001. Review of WAG Field Experience. *SPE Res Eval & Eng* 4 (2): 97–106. SPE-71203-PA. https://doi.org/10.2118/71203-PA.

Clementz, D. M., Patterson, D. E., Aseltine, R. J. et al. 1982. Stimulation of Water Injection Wells in the Los Angeles Basin by Using Sodium Hypochlorite and Mineral Acids. *J Pet Technol* 34 (9): 2087–2096. SPE-10624-PA. https://doi.org/10.2118/10624-PA.

Das, S. C., Fox, G. A., and Zebrowitz, M. J. 1998. Time Lapse Borax Logging in a Karstified Limestone Formation of the Panna Field. Paper presented at the SPE India Oil and Gas Conference and Exhibition, New Delhi, India, 17–19 February. SPE-39544-MS. https://doi.org/10.2118/39544-MS.

Delshad, M., Varavei, A., Goudarzi, A. et al. 2013. Water Management in Mature Oil Fields Using Preformed Particle Gels. Paper presented at the SPE Western Regional & AAPG Pacific Section Meeting 2013 Joint Technical Conference, Monterey, California, USA, 19–25 April. SPE-165356-MS. https://doi.org/10.2118/165356-MS.

Dong, H., Fang, S., Wang, D. et al. 2008. Review of Practical Experience & Management by Polymer Flooding at Daqing. Paper presented at the SPE Symposium on Improved Oil Recovery, Tulsa, Oklahoma, USA, 20–23 April. SPE-114342-MS. https://doi.org/10.2118/114342-MS.

Drummond, A., Fishlock, T., Naylor, P. et al. 2001. An Evaluation of Post-Waterflood Depressurisation of the South Brae Field, North Sea. Paper presented at the SPE Annual Technical Conference and Exhibition, New Orleans, Louisiana, USA, 30 September–3 October. SPE-71487-MS. https://doi.org/10.2118/71487-MS.

Dymond, P. F. and Spurr, P. R. 1988. Magnus Field: Surfactant Stimulation of Water-Injection Wells. SPE Res Eng 3 (1): 165–174. SPE-13980-PA. https://doi.org/10.2118/13980-PA.

Effiom, O. 2016. Achieving Cost Effective 4D Monitoring in Deep Water Fields. Paper presented at the SPE Nigeria Annual International Conference and Exhibition, Lagos, Nigeria, 2–4 August. SPE-184375-MS. https://doi.org/10.2118/184375-MS.

Egermann, P. and Vizika, O. 2000. Critical Gas Saturation and Relative Permeability During Depressurization in the Far Field and the Near-Wellbore Region. Paper presented at the SPE Annual Technical Conference and Exhibition, Dallas, Texas, USA, 1–4 October. SPE-63149-MS. https://doi.org/10.2118/63149-MS.

Elkins, L. F. and Skov, A. M. 1963. Cyclic Water Flooding the Spraberry Utilizes "End Effects" to Increase Oil Production Rate. J Pet Technol 15 (8): 877–884. SPE-545-PA. https://doi.org/10.2118/545-PA.

Eyvazzadeh, R. Y., Kelder, O., Hajari, A. A. et al. 2004. Modern Carbon/Oxygen Logging Methodologies: Comparing Hydrocarbon Saturation Determination Techniques. Paper presented at the SPE Annual Technical Conference and Exhibition, Houston, Texas, USA, 26–29 September. SPE-90339-MS. https://doi.org/10.2118/90339-MS.

Faber, M. J., Joosten, G. J. P., Hashmi, K. A. et al. 1998. Water Shut-Off Field Experience with a Relative Permeability Modification System in the Marmul Field. Paper presented at the SPE/DOE Improved Oil Recovery Symposium, Tulsa, Oklahoma, USA, 19–22 April. SPE-39633-MS. https://doi.org/10.2118/39633-MS.

Faiz, M. F., Mandal, D., Masoudi, R. et al. 2019. Water Injection Performance Benchmarking & Replication of Best Practices Reduces Operating Cost Improves Recovery. Paper presented at the SPE/IATMI Asia Pacific Oil & Gas Conference and Exhibition, Bali, Indonesia, 29–31 October. SPE-196475-MS. https://doi.org/10.2118/196475-MS.

Falls, A. H., Thigpen, D. R., Nelson, R. C. et al. 1994. Field Test of Cosurfactant-Enhanced Alkaline Flooding. SPE Res Eng 9 (3): 217–223. SPE-24117-PA. https://doi.org/10.2118/24117-PA.

Forbes, C. and Taggart, I. 1998. Selective Isolation of Perforated Liners Using Casing Patches: Case Studies from North Sea Operations. Paper presented at the IADC/SPE Drilling Conference, Dallas, Texas, USA, 3–6 March. SPE-39348-MS. https://doi.org/10.2118/39348-MS.

Frampton, H., Morgan, J. C., Cheung, S. K. et al. 2004. Development of a Novel Waterflood Conformance Control System. Paper presented at the SPE/DOE Symposium on Improved Oil Recovery, Tulsa, Oklahoma, USA, 17–21 April. SPE-89391-MS. https://doi.org/10.2118/89391-MS.

Fulleylove, R. J., Morgan, J. C., Stevens, D. G. et al. 1996. Water Shut-Off in Oil Production Wells—Lessons from 12 Treatments. Paper presented at the Abu Dhabi

International Petroleum Exhibition and Conference, Abu Dhabi, UAE, 13–16 October. SPE-36211-MS. https://doi.org/10.2118/36211-MS.

Galiev, A., Mukhliev, I., Volkov, M. et al. 2018. Out of Zone Injection. Paper presented at the SPE Norway One Day Seminar, Bergen, Norway, 18 April. SPE-191338-MS. https://doi.org/10.2118/191338-MS.

Gallagher, J. J., Kemshell, D. M., Taylor, S. R. et al. 1999. Brent Field Depressurization Management. Paper presented at the Offshore Europe Oil and Gas Exhibition and Conference, Aberdeen, UK, 7–10 September. SPE-56973-MS. https://doi.org/10.2118/56973-MS.

Garimella, S. V. S., Al-Kharusi, A. S., Waheibi, H. et al. 2012. Pushing-Up Field Ultimate Recovery to Top Quartile Through Infill Development—A Case Study of a Carbonate Waterflood Field. Paper presented at the Abu Dhabi International Petroleum Conference and Exhibition, Abu Dhabi, UAE, 11–14 November. SPE-161964-MS. https://doi.org/10.2118/161964-MS.

German, M. 2015. Reviewing Waterflood Projects over 12 Years: What Have We Learned? Paper presented at the SPE Kuwait Oil and Gas Show and Conference, Mishref, Kuwait, 11–14 October. SPE-175370-MS. https://doi.org/10.2118/175370-MS.

Giordano, R. M., Jayanti, S., Chopra, A. K. et al. 2007. A Streamline Based Reservoir Management Workflow to Maximize Oil Recovery. Paper presented at the SPE/EAGE Reservoir Characterization and Simulation Conference, Abu Dhabi, UAE, 28–31 October. SPE-111143-MS. https://doi.org/10.2118/111143-MS.

Goodfield, M. and Goodyear, S. G. 2003. Relative Permeabilities for Post-Waterflood Depressurisation. Paper presented at the SPE Offshore Europe Oil and Gas Exhibition and Conference, Aberdeen, UK, 2–5 September. SPE-83958-MS. https://doi.org/10.2118/83958-MS.

Goodwin, N., Walsh, J. M., Wright, R. J. et al. 2011. Modeling the Effect of Triazine Based Sulphide Scavengers on the in situ pH and Scaling Tendency. Paper presented at the SPE International Symposium on Oilfield Chemistry, The Woodlands, Texas, USA, 11–13 April. SPE-141583-MS. https://doi.org/10.2118/141583-MS.

Grinestaff, G. H. and Caffrey, D. J. 2000. Waterflood Management: A Case Study of the Northwest Fault Block Area of Prudhoe Bay, Alaska, Using Streamline Simulation and Traditional Waterflood Analysis. Paper presented at the SPE Annual Technical Conference and Exhibition, Dallas, Texas, USA, 1–4 October. SPE-63152-MS. https://doi.org/10.2118/63152-MS.

Groenenboom, J., Wong, S.-W., Meling, T. et al. 2003. Pulsed Water Injection During Waterflooding. Paper presented at the SPE International Improved Oil Recovery Conference in Asia Pacific, Kuala Lumpur, Malaysia, 20–21 October. SPE-84856-MS. https://doi.org/10.2118/84856-MS.

Guglielmo, D., Soliman, M. Y., Kontarev, R. et al. 2006. Fracture Stimulation in Waterflood Fields of Western Siberia: A Case Study of Water Prediction and Oil Production Optimization Using Multiphase Reservoir Simulation Techniques. Paper presented at the SPE International Symposium and Exhibition on Formation Damage Control, Lafayette, Louisiana, USA, 15–17 February. SPE-98259-MS. https://doi.org/10.2118/98259-MS.

Guo, H., Li, Y., Gu, Y. et al. 2016. Comparison of Strong Alkali and Weak Alkali ASP Flooding Pilot Tests in Daqing Oilfield. Paper presented at the SPE Improved Oil Recovery Conference, Tulsa, Oklahoma, USA, 11–13 April. SPE-179661-MS. https://doi.org/10.2118/179661-MS.

Hadlow, R. E. 1992. Update of Industry Experience with CO2 Injection. Paper presented at the SPE Annual Technical Conference and Exhibition, Washington, DC, USA, 4–7 October. SPE-24928-MS. https://doi.org/10.2118/24928-MS.

Hancock, W. P. 1988. Operating Experience and Expansion of Water Injection Facilities on the Statfjord Field to over 1 Million BWPD. Paper presented at the Offshore Technology Conference, Houston, Texas, USA, 2–5 May. OTC-5839-MS. https://doi.org/10.4043/5839-MS.

Hermansen, H., Thomas, L. K., Sylte, J. E. et al. 1997. Twenty Five Years of Ekofisk Reservoir Management. Paper presented at the SPE Annual Technical Conference and Exhibition, San Antonio, Texas, USA, 5–8 October. SPE-38927-MS. https://doi.org/10.2118/38927-MS.

Hou, J., Liu, Z., and Xia, H. 2001a. Viscoelasticity of ASP Solution Is a More Important Factor of Enhancing Displacement Efficiency than Ultra-Low Interfacial Tension in ASP Flooding. Paper presented at the SPE Rocky Mountain Petroleum Technology Conference, Keystone, Colorado, USA, 21–23 May. SPE-71061-MS. https://doi.org/10.2118/71061-MS.

Hou, J., Liu, Z., Yue, X. et al. 2001b. Study of the Effect of ASP Solution Viscoelasticity on Displacement Efficiency. Paper presented at the SPE Annual Technical Conference and Exhibition, New Orleans, Louisiana, USA, 30 September–3 October. SPE-71492-MS. https://doi.org/10.2118/71492-MS.

Houston, S. J., Yardley, B., Smalley, P. C. et al. 2006. Precipitation and Dissolution of Minerals During a Waterflood—The Evidence of Produced Water Chemistry from Miller. Paper presented at the SPE International Oilfield Scale Symposium, Aberdeen, UK, 31 May–1 June. SPE-100603-MS. https://doi.org/10.2118/100603-MS.

Huseby, O., Clemens, T., Lueftenegger, M. et al. 2016. Analytical Interpretation of Tracer Data to Assess Sweep Changes due to Polymer Injection. Paper presented at IEA's 37th EOR Workshop & Symposium, Rueil-Malmaison, France, 18–22 September.

Hussain, A., Ghufaili, A., and Behera, C. 2019. Urban Planning Creates Value for a Highly Congested Field in Sultanate of Oman. Paper presented at the Abu Dhabi International Petroleum Exhibition & Conference, Abu Dhabi, UAE, 11–14 November. SPE-197225-MS. https://doi.org/10.2118/197225-MS.

Izgec, B. and Kabir, C. S. 2007. Real-Time Performance Analysis of Water-Injection Wells. Paper presented at the SPE Annual Technical Conference and Exhibition, Anaheim, California, USA, 11–14 November. SPE-109876-MS. https://doi.org/10.2118/109876-MS.

Jain, L., Zhang, T., Nguyen, H. et al. 2020. Waterflood Conformance Improvement Method in Naturally Fractured Carbonate Reservoirs with Gel Injection. Paper presented at the International Petroleum Technology Conference, Dhahran, Saudi Arabia, 13–15 January. IPTC-20275-MS. https://doi.org/10.2523/IPTC-20275-MS.

Jalan, S. N., Al-Humoud, J., Al-Sabea, S. H. et al. 2013. Well Integrity: Application of Ultrasonic Logging, Production Logging and Corrosion Logs for Leak Detection in Wells—A Case Study. Paper presented at the SPE Kuwait Oil and Gas Show and Conference, Kuwait City, Kuwait, 8–10 October. SPE-167279-MS. https://doi.org/10.2118/167279-MS.

Jaramillo, O. J., Romero, R., Lucuara, G. et al. 2010. Combining Stimulation and Water Control in High-Water-Cut Wells. Paper presented at the SPE International

Symposium and Exhibition on Formation Damage Control, Lafayette, Louisiana, USA, 10–12 February. SPE-127827-MS. https://doi.org/10.2118/127827-MS.

Jarrell, P. M. and Stein, M. H. 1991. Maximizing Injection Rates in Wells Recently Converted to Injection Using Hearn and Hall Plots. Paper presented at the SPE Production Operations Symposium, Oklahoma City, Oklahoma, USA, 7–9 April. SPE-21724-MS. https://doi.org/10.2118/21724-MS.

Jordan, M. M., Graham, G. M., Sorbie, K. S. et al. 1998. Scale Dissolver Application: Production Enhancement and Formation Damage Potential. Paper presented at the SPE Formation Damage Control Conference, Lafayette, Louisiana, USA, 18–19 February. SPE-39449-MS. https://doi.org/10.2118/39449-MS.

Kabir, A. H., Bakar, M. A., Salim, M. A. et al. 1999. Water/Gas Shut-Off Candidates Selection. Paper presented at the SPE Asia Pacific Oil and Gas Conference and Exhibition, Jakarta, Indonesia, 20–22 April. SPE-54357-MS. https://doi.org/10.2118/54357-MS.

Khanifar, A., Sheykh Alian, S., Demiral, B. et al. 2011. Study of Asphaltene Precipitation and Deposition Phenomenon During WAG Application. Paper presented at the SPE Enhanced Oil Recovery Conference, Kuala Lumpur, Malaysia, 19–21 July. SPE-143488-MS. https://doi.org/10.2118/143488-MS.

Kortukov, D., Shako, V., Pringuey, T. et al. 2019. Fiber Optic Measurements as Real Time PLT with New Transient Interpretation. Paper presented at the SPE/IATMI Asia Pacific Oil & Gas Conference and Exhibition, Bali, Indonesia, 29–31 October. SPE-196272-MS. https://doi.org/10.2118/196272-MS.

Kothiyal, M. D., Kumar, V., Kumar, P. et al. 2012. Improvement of Vertical Conformance in Injectors—A Must for Fluvial Channel Reservoirs. Paper presented at the SPE Oil and Gas India Conference and Exhibition, Mumbai, India, 28–30 March. SPE-155274-MS. https://doi.org/10.2118/155274-MS.

Krebs, T., Balk, W., Verbeek, P. et al. 2016. Debottlenecking of FPSO Facilities by Compact Separators. Paper presented at the Abu Dhabi International Petroleum Exhibition & Conference, Abu Dhabi, UAE, 7–10 November. SPE-183343-MS. https://doi.org/10.2118/183343-MS.

Kume, N. 2003. An Engineering Approach to Predicting Post-Treatment Well Performance Using Selective Chemical Water Shut-Off Techniques: An RPM Example. Paper presented at the SPE Annual Technical Conference and Exhibition, Denver, Colorado, USA, 5–8 October. SPE-84513-MS. https://doi.org/10.2118/84513-MS.

Landa, J. L. and Kumar, D. 2011. Joint Inversion of 4D Seismic and Production Data. Paper presented at the SPE Annual Technical Conference and Exhibition, Denver, Colorado, USA, 30 October–2 November. SPE-146771-MS. https://doi.org/10.2118/146771-MS.

Larsen, J., Zwolle, S., Kjellerup, B. V. et al. 2005. Identification of Bacteria Causing Souring and Biocorrosion in the Halfdan Field by Application of New Molecular Techniques. Paper presented at CORROSION 2005, Houston, Texas, USA, 3–7 April. NACE-05629.

Lee, J. M., Khan, R. I., and Phelps, D. W. 2009. Debottlenecking and Computational-Fluid-Dynamics Studies of High- and Low-Pressure Production Separators. *SPE Proj Fac & Const* 4 (4): 124–131. SPE-115735-PA. https://doi.org/10.2118/115735-PA.

Lerma, M. K. and Giuliani, M. A. 1994. Cost-Effective Methods of Profile Control in Mature Waterflood Injectors. Paper presented at the SPE Western Regional Meeting,

Long Beach, California, USA, 23–25 March. SPE-27849-MS. https://doi.org/10.2118/27849-MS.

Levitt, D., Jackson, A., Heinson, C. et al. 2009. Identification and Evaluation of High-Performance EOR Surfactants. *SPE Res Eval & Eng* 12 (2): 243–253. SPE-100089-PA. https://doi.org/10.2118/100089-PA.

Li, Z. and Zhu, D. 2011. Optimization of Production Performance with ICVs by Using Temperature-Data Feedback in Horizontal Wells. *SPE Prod & Oper* 26 (3): 253–261. SPE-135156-PA. https://doi.org/10.2118/135156-PA.

Ligthelm, D. J., Reijnen, G. C. A. M., Wit, K. et al. 1997. Critical Gas Saturation During Depressurisation and Its Importance in the Brent Field. Paper presented at Offshore Europe, Aberdeen, UK, 9–12 September. SPE-38475-MS. https://doi.org/10.2118/38475-MS.

Ma, T. D. and Youngren, G. K. 1994. Performance of Immiscible Water-Alternating-Gas (IWAG) Injection at Kuparuk River Unit, North Slope, Alaska. Paper presented at the SPE Annual Technical Conference and Exhibition, New Orleans, Louisiana, USA, 25–28 September. SPE-28602-MS. https://doi.org/10.2118/28602-MS.

Mand, J., Jack, T. R., Voordouw, G. et al. 2014. Use of Molecular Methods (Pyrosequencing) for Evaluating MIC Potential in Water Systems for Oil Production in the North Sea. Paper presented at the SPE International Oilfield Corrosion Conference and Exhibition, Aberdeen, UK, 12–13 May. SPE-169638-MS. https://doi.org/10.2118/169638-MS.

Manrique, E., Reyes, S., Romero, J. et al. 2014. Colloidal Dispersion Gels (CDG): Field Projects Review. Paper presented at the SPE EOR Conference at Oil and Gas West Asia, Muscat, Oman, 31 March–2 April. SPE-169705-MS. https://doi.org/10.2118/169705-MS.

Matthew, J., Harmon, C. E., Harmon, S. C. et al. 2004. New Expandable Cladding Technique Enables Extended Length Casing Repair. Paper presented at the IADC/SPE Drilling Conference, Dallas, Texas, USA, 2–4 March. SPE-87212-MS. https://doi.org/10.2118/87212-MS.

Maxwell, S. 2007. Recent Advances in the Application of Nitrate to Control Reservoir Souring. Paper presented at the International Symposium on Oilfield Chemistry, Houston, Texas, USA, 28 February–2 March. SPE-106467-MS. https://doi.org/10.2118/106467-MS.

Maxwell, S., Devine, C., and Rooney, F. 2004. Monitoring and Control of Bacterial Biofilms in Oilfield Water Handling Systems. Paper presented at CORROSION 2004, New Orleans, Louisiana, USA, 28 March–1 April. NACE-04752.

Maxwell, S., Mutch, K., Charlton, P. et al. 2002. In-Field Biocide Optimization for Magnus Water Injection Systems. Paper presented at CORROSION 2002, Denver, Colorado, USA, 7–11 April. NACE-02031.

Mayer, E. H., Berg, R. L., Carmichael, J. D. et al. 1983. Alkaline Injection for Enhanced Oil Recovery—A Status Report. *J Pet Technol* 35 (1): 209–221. SPE-8848-PA. https://doi.org/10.2118/8848-PA.

Medd, D. M., Thomas, P., Sibbons, C. et al. 2010. Integration of 4D Seismic to Add Value: The Enfield ENC01 Sidetrack Story. Paper presented at the SPE Asia Pacific Oil and Gas Conference and Exhibition, Brisbane, Queensland, Australia, 18–20 October. SPE-136538-MS. https://doi.org/10.2118/136538-MS.

Medeiros, R. S., Deepankar, B., and Suryanarayana, P. V. 2004. Impact of Thief Zone Identification and Shut-Off on Water Production in the Nimr Field. Paper presented at the SPE/IADC Underbalanced Technology Conference and Exhibition, Houston, Texas, USA, 11–12 October. SPE-91665-MS. https://doi.org/10.2118/91665-MS.

Mijnarends, R., Frolov, A., Grishko, F. et al. 2015. Advanced Data-Driven Performance Analysis for Mature Waterfloods. Paper presented at the SPE Annual Technical Conference and Exhibition, Houston, Texas, USA, 28–30 September. SPE-174872-MS. https://doi.org/10.2118/174872-MS.

Mikkelsen, P. L., Guderian, K., and du Plessis, G. 2005. Improved Reservoir Management Through Integration of 4D Seismic Interpretation, Draugen Field, Norway. Paper presented at the SPE Offshore Europe Oil and Gas Exhibition and Conference, Aberdeen, UK, 6–9 September. SPE-96400-MS. https://doi.org/10.2118/96400-MS.

Mishra, S., Shrivastava, C., Ojha, A. et al. 2019. Waterflood Surveillance by Calibrating Streamline-Based Simulation with Crosswell Electromagnetic Data. Paper presented at the International Petroleum Technology Conference, Beijing, China, 26–28 March. IPTC-19286-MS. https://doi.org/10.2523/IPTC-19286-MS.

Moffitt, P. D., Zornes, D. R., Moradi-Araghi, A. et al. 1993. Application of Freshwater and Brine Polymer Flooding in the North Burbank Unit, Osage County, Oklahoma. *SPE Res Eng* **8** (2): 128–134. SPE-20466-PA. https://doi.org/10.2118/20466-PA.

Mohammed, M. A., Al-Mubarak, M. A., Al-Mulhim, A. K. et al. 1998. Overview of Field Results Using TTBP as Effective Water Shut-Off Treatment in Openhole Well Completions in Saudi Arabia. Paper presented at the SPE/DOE Improved Oil Recovery Symposium, Tulsa, Oklahoma, USA, 19–22 April. SPE-39616-MS. https://doi.org/10.2118/39616-MS.

Montes, A., Nyhavn, F., Oftedal, G. et al. 2013. Application of Inflow Well Tracers for Permanent Reservoir Monitoring in North Amethyst Subsea Tieback ICD Wells in Canada. Paper presented at the SPE Middle East Intelligent Energy Conference and Exhibition, Manama, Bahrain, 28–30 October. SPE-167463-MS. https://doi.org/10.2118/167463-MS.

Mustoni, J. L., Denyer, P., and Norman, C. 2010. Deep Conformance Control by a Novel Thermally Activated Particle System to Improve Sweep Efficiency in Mature Waterfloods of the San Jorge Basin. Paper presented at the SPE Improved Oil Recovery Symposium, Tulsa, Oklahoma, USA, 24–28 April. SPE-129732-MS. https://doi.org/10.2118/129732-MS.

Naguib, M. A., Sikaiti, S., Balushi, H. et al. 2006. Results of Proactively Managing a Heavy-Oil Waterflood in South Oman Using Streamline-Based Simulation. Paper presented at the SPE Asia Pacific Oil & Gas Conference and Exhibition, Adelaide, Australia, 11–13 September. SPE-101195-MS. https://doi.org/10.2118/101195-MS.

Nakayama, S., Belaid, K., and Ishiyama, T. 2012. 3D OBC Seismic Data Acquisition Productivity Enhancement. Paper presented at the Abu Dhabi International Petroleum Conference and Exhibition, Abu Dhabi, UAE, 11–14 November. SPE-160944-MS. https://doi.org/10.2118/160944-MS.

Nelson, R. C., Lawson, J. B., Thigpen, D. R. et al. 1984. Cosurfactant-Enhanced Alkaline Flooding. Paper presented at the SPE Enhanced Oil Recovery Symposium, Tulsa, Oklahoma, 15–18 April. SPE-12672-MS. https://doi.org/10.2118/12672-MS.

Nikjoo, E. 2019. Novel Realtime Tracer Technology for Continuous Well and Reservoir Monitoring. Paper presented at the Abu Dhabi International Petroleum

Conference and Exhibition, Abu Dhabi, UAE, 11–14 November. SPE-197691-MS. https://doi.org/10.2118/197691-MS.

Nitzberg, K. E. and Broman, W. H. 1992. Improved Reservoir Characterization from Waterflood Tracer Movement, Northwest Fault Block, Prudhoe Bay, Alaska. *SPE Form Eval* 7 (3): 228–234. SPE-20548-PA. https://doi.org/10.2118/20548-PA.

Noll, L. A. 1991. The Effect of Temperature, Salinity, and Alcohol on the Critical Micelle Concentration of Surfactants. Paper presented at the SPE International Symposium on Oilfield Chemistry, Anaheim, California, USA, 20–22 February. SPE-21032-MS. https://doi.org/10.2118/21032-MS.

Nwogu, I. C., Ayo, A., Asemota, O. et al. 2019. Successful Application of Capacitance Resistance Modeling to Understand Reservoir Dynamics in Brown Field Waterflood; A Niger Delta Swamp Field Case Study. Paper presented at the SPE Nigeria Annual International Conference and Exhibition, Lagos, Nigeria, 5–7 August. SPE-198819-MS. https://doi.org/10.2118/198819-MS.

Ohms, D., McLeod, J. D., Graff, C. J. et al. 2010. Incremental-Oil Success from Waterflood Sweep Improvement in Alaska. *SPE Prod & Oper* 25 (3): 247–254. SPE-121761-PA. https://doi.org/10.2118/121761-PA.

Onwuchekwa, V., Usman, M., Wantong, P. et al. 2019. Injectivity Monitoring & Evolution for Water Injectors in a Deepwater Turbidite Field. Paper presented at the SPE Nigeria Annual International Conference and Exhibition, Lagos, Nigeria, 5–7 August. SPE-198749-MS. https://doi.org/10.2118/198749-MS.

Oosthuizen, U. R., Al-Naqi, A. M. H., Al-Anzi, K. et al. 2007. Horizontal Well Production Logging Experience in Heavy Oil Environment with Sand Screen: A Case Study from Kuwait. Paper presented at the SPE Middle East Oil and Gas Show and Conference, Manama, Bahrain, 11–14 March. SPE-105327-MS. https://doi.org/10.2118/105327-MS.

Pajonk, O., Schulze-Riegert, R., Krosche, M. et al. 2011. Ensemble-Based Water Flooding Optimization Applied to Mature Fields. Paper presented at the SPE Middle East Oil and Gas Show and Conference, Manama, Bahrain, 25–28 September. SPE-142621-MS. https://doi.org/10.2118/142621-MS.

Pandey, A., Suresh Kumar, M., Jha, M. K. et al. 2012. Chemical EOR Pilot in Mangala Field: Results of Initial Polymer Flood Phase. Paper presented at the SPE Improved Oil Recovery Symposium, Tulsa, Oklahoma, USA, 14–18 April. SPE-154159-MS. https://doi.org/10.2118/154159-MS.

Paulo, J., Mackay, E. J., Menzies, N. et al. 2001. Implications of Brine Mixing in the Reservoir for Scale Management in the Alba Field. Paper presented at the International Symposium on Oilfield Scale, Aberdeen, UK, 30–31 January. SPE-68310-MS. https://doi.org/10.2118/68310-MS.

Perez, D., Salicioni, F., and Ucan, S. 2014. Cyclic Water Injection in San Jorge Gulf Basin, Argentina. Paper presented at the SPE Latin America and Caribbean Petroleum Engineering Conference, Maracaibo, Venezuela, 21–23 May. SPE-169403-MS. https://doi.org/10.2118/169403-MS.

Plante, M. E. and Mackenzie, G. R. J. 2000. Selective Chemical Water Shutoffs Utilizing Through-Tubing Inflatable Packer Technology. Paper presented at the SPE/ICoTA Coiled Tubing Roundtable, Houston, Texas, USA, 5–6 April. SPE-60717-MS. https://doi.org/10.2118/60717-MS.

Portwood, J. T. 1999. Lessons Learned from over 300 Producing Well Water Shut-Off Gel Treatments. Paper presented at the SPE Mid-Continent Operations

Symposium, Oklahoma City, Oklahoma, USA, 28–31 March. SPE-52127-MS. https://doi.org/10.2118/52127-MS.

Prelicz, R. M., Fearfield, D., Sobera, M. et al. 2014. Identifying New Opportunities Through Reservoir Performance Reviews and Systematic Benchmarking of TQ Recovery. Paper presented at the Abu Dhabi International Petroleum Exhibition and Conference, Abu Dhabi, UAE, 10–13 November. SPE-172095-MS. https://doi.org/10.2118/172095-MS.

Pritchett, J., Frampton, H., Brinkman, J. et al. 2003. Field Application of a New In-Depth Waterflood Conformance Improvement Tool. Paper presented at the SPE International Improved Oil Recovery Conference in Asia Pacific, Kuala Lumpur, Malaysia, 20–21 October. SPE-84897-MS. https://doi.org/10.2118/84897-MS.

Qi, Q., Pepin, S., AlJazzaf, A. et al. 2017. Errors Associated with Waterflood Monitoring Using the Hall Plot for Stacked Reservoirs in the Absence of Profile Surveys. Paper presented at the SPE Western Regional Meeting, Bakersfield, California, USA, 23–27 April. SPE-185694-MS. https://doi.org/10.2118/185694-MS.

Rahman, M., Zannitto, P. J., Reed, D. A. et al. 2011. Application of Fiber-Optic Distributed Temperature Sensing Technology for Monitoring Injection Profile in Belridge Field, Diatomite Reservoir. Paper presented at the SPE Digital Energy Conference and Exhibition, The Woodlands, Texas, USA, 19–21 April. SPE-84897-MS. https://doi.org/10.2118/144116-MS.

Rajamani, K. and Ipsen, M. 2016. Efficient Installation of a Modular Straddle System for Deep Water Gas or Water Shut Offs. Paper presented at the SPE Nigeria Annual International Conference and Exhibition, Lagos, Nigeria, 2–4 August. SPE-184337-MS. https://doi.org/10.2118/184337-MS.

Rawlins, C. H. 2017. Partial Processing: Produced Water Debottlenecking Unlocks Production on Offshore Thailand MOPU Platform. Paper presented at the SPE Annual Technical Conference and Exhibition, San Antonio, Texas, USA, 9–11 October. SPE-187109-MS. https://doi.org/10.2118/187109-MS.

Retail, P., Thane, L., and Liberelle, E. 2002. Well Interference Test on Girassol. Paper presented at the Offshore Technology Conference, Houston, Texas, USA, 6–9 May. OTC-14167-MS. https://doi.org/10.4043/14167-MS.

Reviere, R. H. and Wu, C. H. 1986. An Economic Evaluation of Waterflood Infill Drilling in Nine Texas Waterflood Units. Paper presented at the Permian Basin Oil and Gas Recovery Conference, Midland, Texas, USA, 13–15 March. SPE-15037-MS. https://doi.org/10.2118/15037-MS.

Roberts, J. A. and Roberts, R. S. 2005. A Novel Approach to Eliminating Sulfur Deposition in Liquid Redox Hydrogen Sulfide Removal Systems. Paper presented at the SPE Western Regional Meeting, Irvine, California, USA, 30 March–1 April. SPE-93841-MS. https://doi.org/10.2118/93841-MS.

Roussennac, B. D. and Toschi, C. 2010. Brightwater Trial in Salema Field (Campos Basin, Brazil). Paper presented at the SPE EUROPEC/EAGE Annual Conference and Exhibition, Barcelona, Spain, 14–17 June. SPE-131299-MS. https://doi.org/10.2118/131299-MS.

Rublev, A. B., Khuzeev, Y. A., Ishimov, I. A. et al. 2012. Predictions of Cyclic Water Injection on Urnenskoe Oil Field. Paper presented at the SPE Russian Oil and Gas Exploration and Production Technical Conference and Exhibition, Moscow, Russia, 16–18 October. SPE-162015-MS. https://doi.org/10.2118/162015-MS.

Sayarpour, M., Kabir, C. S., and Lake, L. W. 2008. Field Applications of Capacitance Resistive Models in Waterfloods. Paper presented at the SPE Annual Technical Conference and Exhibition, Denver, Colorado, USA, 21–24 September. SPE-114983-MS. https://doi.org/10.2118/114983-MS.

Scheck, M. and Ross, G. 2008. Improvement of Scale Management Using Analytical and Statistical Tools. Paper presented at the SPE International Oilfield Scale Conference, Aberdeen, UK, 28–29 May. SPE-114103-MS. https://doi.org/10.2118/114103-MS.

Seright, R. S., Zhang, G., Akanni, O. O. et al. 2011. A Comparison of Polymer Flooding with In-Depth Profile Modification. Paper presented at the Canadian Unconventional Resources Conference, Calgary, Alberta, Canada, 15–17 November. SPE-146087-MS. https://doi.org/10.2118/146087-MS.

Shchipanov, A., Surguchev, L. M., and Jakobsen, S. R. 2008. Improved Oil Recovery by Cyclic Injection and Production. Paper presented at the SPE Russian Oil and Gas Technical Conference and Exhibition, Moscow, Russia, 28–30 October. SPE-116873-MS. https://doi.org/10.2118/116873-MS.

Shnaib, F. Y., Desouky, A. M. M., Mehrotra, N. et al. 2009. Case Study of Successful Matrix Stimulation of High-Water-Cut Wells in Dubai Offshore Fields. Paper presented at the International Petroleum Technology Conference, Doha, Qatar, 7–9 December. IPTC-13203-MS. https://doi.org/10.2523/IPTC-13203-MS.

Shook, G. M., Pope, G. A., and Asakawa, K. 2009. Determining Reservoir Properties and Flood Performance from Tracer Test Analysis. Paper presented at the SPE Annual Technical Conference and Exhibition, New Orleans, Louisiana, USA, 4–7 October. SPE-124614-MS. https://doi.org/10.2118/124614-MS.

Shutang, G. and Qiang, G. 2010. Recent Progress and Evaluation of ASP Flooding for EOR in Daqing Oil Field. Paper presented at the SPE EOR Conference at Oil & Gas West Asia, Muscat, Oman, 11–13 April. SPE-127714-MS. https://doi.org/10.2118/127714-MS.

Sibilev, V., Tokareva, O., Kolesov, V. et al. 2019. 4D TEM Surveys for Waterflood Monitoring in a Carbonate Reservoir. Paper presented at the SPE Russian Petroleum Technology Conference, Moscow, Russia, 22–24 October. SPE-196943-MS. https://doi.org/10.2118/196943-MS.

Skovhus, T. L., Hojris, B., Saunders, A. M. et al. 2007. Practical Use of New Microbiology Tools in Oil Production. Paper presented at the SPE Offshore Europe Oil and Gas Conference and Exhibition, Aberdeen, UK, 4–7 September. SPE-109104-MS. https://doi.org/10.2118/109104-MS.

Skrettingland, K., Dale, E. I., Stenerud, V. R. et al. 2014. Snorre In-Depth Water Diversion Using Sodium Silicate—Large Scale Interwell Field Pilot. Paper presented at the SPE EOR Conference at Oil and Gas West Asia, Muscat, Oman, 31 March–2 April. SPE-169727-MS. https://doi.org/10.2118/169727-MS.

Sommer, F. S. and Jenkins, D. P. 1993. Channel Detection Using Pulsed Neutron Logging in a Borax Solution. Paper presented at the SPE Asia Pacific Oil and Gas Conference, Singapore, 8–10 February. SPE-25383-MS. https://doi.org/10.2118/25383-MS.

Stammeijer, J. G. F., Davidson, M., Hatchell, P. J. et al. 2013. Instantaneous 4D Seismic (i4D)—An Innovative Concept to Monitor Offshore Water Injector Wells. Paper presented at the International Petroleum Technology Conference, Beijing, China, 26–28 March. IPTC-16901-MS. https://doi.org/10.2523/IPTC-16901-MS.

Starcher, M., Mut, D., Fontanarosa, M. et al. 2002. Next Generation Waterflood Surveillance: Behind Casing Resistivity Measurement Successfully Applied in the 'A3-A6' Waterflood at Elk Hills Field, Kern County, California. Paper presented at the SPE Western Regional/AAPG Pacific Section Joint Meeting, Anchorage, Alaska, USA, 20–22 May. SPE-76730-MS. https://doi.org/10.2118/76730-MS.

Stavland, A., Jonsbraten, H. C., Vikane, O. et al. 2011. In-Depth Water Diversion Using Sodium Silicate on Snorre—Factors Controlling In-Depth Placement. Paper presented at the SPE European Formation Damage Conference, Noordwijk, Netherlands, 7–10 June. SPE-143836-MS. https://doi.org/10.2118/143836-MS.

Steagall, D. E., Gomes, J. A. T., De Oliveira, R. M. et al. 2005. How to Estimate the Value of the Information (VOI) of a 4D Seismic Survey in One Offshore Giant Field. Paper presented at the SPE Annual Technical Conference and Exhibition, Dallas, Texas, USA, 9–12 October. SPE-95876-MS. https://doi.org/10.2118/95876-MS.

Surguchev, L. M., Giske, N. H., Kollbotn, L. et al. 2008. Cyclic Water Injection Improves Oil Production in Carbonate Reservoir. Paper presented at the Abu Dhabi International Petroleum Exhibition and Conference, Abu Dhabi, UAE, 3–6 November. SPE-117836-MS. https://doi.org/10.2118/117836-MS.

Surguchev, L. M., Korbol, R., Haugen, S. et al. 1992. Screening of WAG Injection Strategies for Heterogeneous Reservoirs. Paper presented at the European Petroleum Conference, Cannes, France, 16–18 November. SPE-25075-MS. https://doi.org/10.2118/25075-MS.

Surguchev, L. M., Virnovsky, G., Reich, E-M. et al. 2003. Evaluation of Cyclic Waterflooding and IOR Screening. Paper presented at IOR 2003—12th European Symposium on Improved Oil Recovery, September. cp-7-00029. https://doi.org/10.3997/2214-4609-pdb.7.A031.

Sydansk, R. D. and Seright, R. S. 2006. When and Where Relative Permeability Modification Water-Shutoff Treatments Can Be Successfully Applied. Paper presented at the SPE/DOE Symposium on Improved Oil Recovery, Tulsa, Oklahoma, USA, 22–26 April. SPE-99371-MS. https://doi.org/10.2118/99371-MS.

Taylor, G. N., Wylde, J. J., Müller, T. et al. 2017. Fresh Insight into the H_2S Scavenging Mechanism of MEA-Triazine vs. MMA-Triazine. Paper presented at the SPE International Conference on Oilfield Chemistry, Montgomery, Texas, USA, 3–5 April. SPE-184529-MS. https://doi.org/10.2118/184529-MS.

Tayyib, D., Al-Qasim, A., Kokal, S. et al. 2019. Overview of Tracer Applications in Oil and Gas Industry. Paper presented at the SPE Kuwait Oil & Gas Show and Conference, Mishref, Kuwait, 13–16 October. SPE-198157-MS. https://doi.org/10.2118/198157-MS.

Tehrani, A. D. H., Smart, B. G. D., and Shwishin, N. M. M. 2002. Importance of Compaction in Depressurisation of Oil Reservoirs. Paper presented at the SPE Asia Pacific Oil and Gas Conference and Exhibition, Melbourne, Australia, 8–10 October. SPE-77968-MS. https://doi.org/10.2118/77968-MS.

Terrado, R. M., Yudono, S., and Thakur, G. C. 2007. Waterflooding Surveillance and Monitoring: Putting Principles into Practice. *SPE Res Eval & Eng* **10** (5): 552–562. SPE-102200-PA. https://doi.org/10.2118/102200-PA.

Turta, A. T. and Singhal, A. K. 2000. Overview of Short Distance Oil Displacement Processes. Paper presented at the SPE/CIM International Conference on Horizontal Well Technology, Calgary, Alberta, Canada, 6–8 November. SPE-66791-MS. https://doi.org/10.2118/66791-MS.

van Eijden, J., Arkesteijn, F., Akil, I. et al. 2004. Gel-Cement, a Water Shut-Off System: Qualification in a Syrian Field. Paper presented at the Abu Dhabi International Conference and Exhibition, Abu Dhabi, UAE, 10–13 October. SPE-88765-MS. https://doi.org/10.2118/88765-MS.

van Everdingen, A. F. and Kriss, H. S. 1980. A Proposal to Improve Recovery Efficiency. *J Pet Technol* **32** (7): 1164–1168. SPE-9088-PA. https://doi.org/10.2118/9088-PA.

van Poelgeest, F., Niko, H., and Medwid, A. R. 1991. Comparison of Laboratory and In-Situ Measurements of Waterflood Residual Oil Saturations for the Cormorant Field. *SPE Form Eval* **6** (1): 39–44. SPE-19300-PA. https://doi.org/10.2118/19300-PA.

van Poolen, H. K. 1980. *Fundamentals of Enhanced Oil Recovery*. Tulsa, Oklahoma: PennWell Books.

Viig, S. O., Juilla, H., Renouf, P. et al. 2013. Application of a New Class of Chemical Tracers to Measure Oil Saturation in Partitioning Interwell Tracer Tests. Paper presented at the SPE International Symposium on Oilfield Chemistry, The Woodlands, Texas, USA, 8–10 April. SPE-164059-MS. https://doi.org/10.2118/164059-MS.

Vuori, V., Nuutinen, V., Honkanen, T. et al. 2016. Accurate Detection of Tagged Polymeric Scale Inhibitors in Oilfield Produced Water Samples. Paper presented at the SPE International Oilfield Scale Conference and Exhibition, Aberdeen, UK, 11–12 May. SPE-179908-MS. https://doi.org/10.2118/179908-MS.

Walker, T., Kerns, S., Scott, D. et al. 2002. Fracture Stimulation Optimization in the Redevelopment of a Mature Waterflood, Elk Hills Field, California. Paper presented at the SPE Western Regional/AAPG Pacific Section Joint Meeting, Anchorage, Alaska, USA, 20–22 May. SPE-76723-MS. https://doi.org/10.2118/76723-MS.

Wan, X., Huang, M., Yu, H. et al. 2006. Pilot Test of Weak Alkaline System ASP Flooding in Secondary Layer with Small Well Spacing. Paper presented at the International Oil & Gas Conference and Exhibition in China, Beijing, China, 5–7 December. SPE-104416-MS. https://doi.org/10.2118/104416-MS.

Warner, H. R., Jr. 2015. *The Reservoir Engineering Aspects of Waterflooding*, second edition, Vol. 3. Richardson, Texas: Monograph Series, Society of Petroleum Engineers.

Wayhan, D. A. 1972. The Effect of Operating Decisions on Peripheral Waterflood Cash Flow. *J Pet Technol* **24** (11): 1320–1324. SPE-3688-PA. https://doi.org/10.2118/3688-PA.

Webb, P. J. and Kuhn, O. 2004. Enhanced Scale Management Through the Application of Inorganic Geochemistry and Statistics. Paper presented at the SPE International Symposium on Oilfield Scale, Aberdeen, UK, 26–27 May. SPE-87458-MS. https://doi.org/10.2118/87458-MS.

Welling, R. W., Marketz, F., Moosa, R. et al. 2007. Inflow Profile Control in Horizontal Wells in a Fractured Carbonate Using Swellable Elastomers. Paper presented at the SPE Middle East Oil and Gas Show and Conference, Manama, Bahrain, 11–14 March. SPE-105709-MS. https://doi.org/10.2118/105709-MS.

Whitesell, L. B. 1961. Field Evaluation of Microbial Problems and Their Control. Paper presented at the Annual Meeting of Rocky Mountain Petroleum Engineers of AIME, Farmington, New Mexico, USA, 26–27 May. SPE-64-MS. https://doi.org/10.2118/64-MS.

Willhite, G. P. 1986. *Waterflooding*, Vol. 3. Richardson, Texas: Textbook Series, Society of Petroleum Engineers.

Williams, H., Dyer, S., Bezerra, M. C. M. et al. 2010. The Effect of Sulphide Scavengers on Scaling Tendency and Scale Inhibitor Performance. Paper presented at the SPE International Conference on Oilfield Scale, Aberdeen, UK, 26–27 May. SPE-131115-MS. https://doi.org/10.2118/131115-MS.

Wu, J. and Ershaghi, I. 2019. Saturation Dependent Errors in Interpreting Waterflood Diagnostic Plots. Paper presented at the SPE Western Regional Meeting, San Jose, California, USA, 23–26 April. SPE-195349-MS. https://doi.org/10.2118/195349-MS.

Wu, X., Zhang, G., Wang, H. et al. 2008. Application of Surfactants with Narrow Equivalent Weight Distribution and Desirable Structure to Daqing ASP Flooding. Paper presented at the SPE Symposium on Improved Oil Recovery, Tulsa, Oklahoma, USA, 20–23 April. SPE-114345-MS. https://doi.org/10.2118/114345-MS.

Yang, Z. 2008. A New Diagnostic Analysis Method for Waterflood Performance. Paper presented at the SPE Western Regional and Pacific Section AAPG Joint Meeting, Bakersfield, California, USA, 29 March–4 April. SPE-113856-MS. https://doi.org/10.2118/113856-MS.

Yu, T., Lei, Z., Li, J. et al. 2018. Infill Drilling Optimization in Waterflooded Tight-Low Permeability Reservoir. Paper presented at the SPE Kingdom of Saudi Arabia Annual Technical Symposium and Exhibition, Dammam, Saudi Arabia, 23–26 April. SPE-192416-MS. https://doi.org/10.2118/192416-MS.

Zhang, Y., Daniels, J. K., Hardy-Fidoe, J. et al. 2014. Scale Inhibitor Residual Analysis: Twenty-First Century Approach. Paper presented at the SPE International Oilfield Scale Conference and Exhibition, Aberdeen, UK, 14–15 May. SPE-169773-MS. https://doi.org/10.2118/169773-MS.

Zubarev, D., Mardanov, R., Bochkarev, V. A. et al. 2019. Flexible Multi-Well Interference Test Design for a Deep-Water Field. Paper presented at the SPE Russian Petroleum Technology Conference, Moscow, Russia, 22–24 October. SPE-196837-MS. https://doi.org/10.2118/196837-MS.

SI Metric Conversion Factors

bbl $\times$ 1.589 873 E–01 = m^3

°F (°F – 32)/1.8 = °C

PetroBriefs

SPE's newest book series is meant to quickly bring each reader up to speed on an emerging technology or specialized topic. At approximately 100 pages, most PetroBriefs are available in both softcover and popular eBook formats.

To see a current list of available PetroBriefs, visit store.spe.org.

Authored by Dave Chappell

Waterflooding: Chemistry
Waterflooding: Facilities and Operations
Waterflooding: Design and Development
Waterflooding: Surveillance and Remediation
Waterflooding: Injection Regime and Injection Wells

Society of Petroleum Engineers

SPE is the recognized leader for publications in the upstream oil and gas industry, and the SPE Bookstore is your source for the books that set the standards of excellence.

Don't miss out on the latest from SPE. Sign up to receive information about new releases at store.spe.org.

www.ingramcontent.com/pod-product-compliance
Lightning Source LLC
Chambersburg PA
CBHW042311210326
41598CB00041B/7351